National Library of Australia.
Cataloguing in Australia.
Seears, Russell John

"Practical Beef Cow Management"

ISBN-13: 978-1481178372

ISBN-10: 1481178377

© **Copyright:** 2011. All rights reserved. No part of this book can be copied without written permission from the author.

Mobile: +61 (0) 428780212
Email: seears@activ8.net.au
Address: Australia

ABN: 5529 2543 792.

Set up by Russell Seears.

Disclaimer:

The contents of this book is information gained from the raw practical experience of the author over many years in the Australian Cattle Industry as a Jackaroo, Ringer/Stockman, Beef Cattle Artificial Inseminator and Cattle Property Manager. All the information, ideas and suggestions for the management of a beef cattle breeding herd, are solely the views of the author and have not been sourced from other written information, except for the photos in the section on Mating Weights, on pages, 106 – 112. They were sourced from and with the authority of, the New South Wales Department of Agriculture, NSW Australia.

Acknowledgements:

I want to thank Barbara Walthall of "Tavinghi" and Jim Connolly of "Overdale", Queensland Australia, for their kindness in allowing me to use photos of their properties and cattle, for the purpose of illustrating a view point in this book.

Page - Contents

4. Introduction
7. General Planning
7. a. End product
10. b. Choosing the cattle breed
14. c. Paddock design
25. d. Watering points
39. e. Pasture requirements
49. f. Yard design and drafting cattle
61. Mustering and moving cattle
72. Weaning
102. Growing and selection period
103. Managing natural mating
103. a. Mating weights
113. b. Herd health
115. c. Oestrus observation
118. d. Heifers mating age
122. Bull selection
130. Paddock mating system
135. Artificial breeding program
146. Monitoring cows from joining to calving
149. Carrying out calving duties
155. Calf marking
161. Weaning
162. Summary

Introduction

This book deals with ideas, information and procedures based on the experiences of the author. It covers the day to day practical management and handling of a beef cattle breeding herd under open range conditions.

The information on each topic has been broadened to include a wide reading audience and will be easily understood by all readers with different levels of rural experience and knowledge. This is not a technical book, but rather a book that is based on, and looks at, the practical day to day management of breeding cows.

The author has had forty years of hands-on experience with breeding cows in many states of Australia, on herds up to 3,000 cows and on cattle properties from 600 acres up to 300,000 acres.

These cattle properties range from small acreages in high rainfall environments, to large cattle stations in the tropical and semi-arid areas of central and northern Australia.

The author recognizes and promotes the principle, that cattle farmers always select or purchase breeding cows and bulls with positive breeding traits, such as temperament, fertility, calving ease, birth weights, milking ability, weaning weights and weight gains etc. However, this book will outline how different management skills and procedures can achieve most of those positive breeding traits through natural selection and management.

The result will be a higher quality breeding herd, in terms of increased herd fertility, higher weaning weights and weight gains and more efficient and profitable breeding cows.

Farmers will see how better handling practices and getting to know and understand cattle and their habits, can make the daily management of their herd run more smoothly, with less stress when mustering and drafting cattle in the yards. This will result in a higher quality end-product and increased profit margin at time of sale.

The first chapter on General Planning will start in the beginning, looking from a different point of view at what the end product of a cattle breeding program is. It will point out how a different approach to cattle breeding and their handling can increase the profitability of a farmer's business and industry.

The chapter will also discuss certain management practices and how these can help make this different point of view, workable and rewarding.

This book will then look at a group of weaners, examining the method of weaning and how their mothers can be selected as breeders by the quality of their calves.

It will then follow the growing period of the heifer weaners only, through to when they are mature, mated, have their calves on the ground and those calves are ready to be weaned.

General Planning

This chapter is for farmers who are planning to start out in the cattle industry, as well as those who would like to make improvements to their breeding and management systems, their end product or the quality of their breeding herd in their current operations. It will also make the information in this book more easily understood, as reasons will be given to why certain management or handling procedures are being suggested.

The following information will be broken down into sections in order of thought processes, to aid in the understanding of the author's intentions.

a. End Product:

This is probably the most important section of this book and needs to be addressed by all farmers before a breeding program begins.

- Farmers should give some thought to the reason WHY they are breeding cattle and for what purpose?
- They should examine what will be the

result and final product of their cattle-breeding program.

- What market will be targeted and what is the required weight, age and quality of the end product for that market?

- What cost is involved to reach that market?

- What is the cost/profit ratio, are the returns justified?

To do this successfully, most farmers may need to have a complete change of thought and mindset to cattle breeding and their handling. Whatever breed of cattle is chosen and whatever type of market is targeted, the end result is meat production, and meat is FOOD, so in reality cattle farmers are producing FOOD.

With this in mind, farmers should be aware at all times during the breeding program and whenever cattle are being handled or a herd management system is being planned, that they are producing and handling FOOD for human consumption.

To emphasise this change of thinking to cattle breeding and its end product, farmers need only to consider what they look for when purchasing other fresh food items, such as fruit and vegetables.

Consumers look for appearance, colour, flavour and texture, chemical free and palatable when cooked. One never sees a farmer hit a cabbage with a piece of poly pipe or stick, or give it a jab with an electric jigger, or yell and curse at it before he takes the cabbage up to the check-out counter to be purchased, so WHY treat cattle that way? It is all food.

So when designing the layout of paddocks and yards, mustering or handling cattle, pasture and grazing requirements, culling rejects for poor breeder quality, conformation or temperament, always remember that your management and handling practices are governing your end product to consumers. These practices are also having an influence on the farmer's profitability as a cattle breeder and FOOD producer.

b. Choosing the Cattle Breed:

Before farmers embark on a cattle breeding program, they may want to look at what breed and type of cattle will suit their area, taking into consideration the following points;

- The type of country.

- The size of the property eg, distance to water for the cattle on a daily basis.

- External and internal parasites such as cattle ticks, buffalo fly, lice, worms etc.

- Rainfall and whether the rainfall is in summer or winter and what is the yearly average.

- Weather temperatures.

- Closest markets.

- Cost of labour.

- Capital available etc.

The reason why this research is so critical to the success of a breeding program is for example, in Northern Australia because of the size of the properties and the environment, there would not be a cattle industry if it were not for the Bos Indicus breeds of cattle.eg, Brahmans.

Approximately 75% of all northern cattle are infused with a percentage of Bos Indicus blood.

From this information it can be concluded, that a percentage of Bos Indicus blood is very beneficial, if not essential, to the survival and prosperity of cattle farmers in Northern Australia.

In the southern states, this percentage is a lot less and in many herds, there is no Bos Indicus blood, just pure British or European breeds or a mixture of the two. The smaller size of the properties and the softer environment suits these breeds more favourably.

This is where farmers have to make decisions with their heads and not their hearts. Breeding cattle is a business, so treat it that way, select a breed of cattle that will give you a profitable and rewarding cattle breeding business.

To obtain this information, farmers could contact their local Stock Agent, Department of Primary Industries, Department of Agricultural, local Primary Industry Consultants, or talk to other cattle breeders in their area, especially those who have been breeding cattle for a long period in the same area.

You may find that a crossbred animal will give you the best results in your area for producing your end product. Keep your mind open to all possibilities, experiment a little with different breeds of cattle, cross one with another and see what you get.

Charbray Breed of beef cattle, average 70% Brahman plus 30% Charolais.

When a decision has been made on the breed of cattle to use, learn as much about that breed as possible, for example;

- What is its origin?

- What is the breed best used for, is it best suited to producing vealers or twelve to twenty-four month old fat steers?

- Will the breed give you the product you require for your available market?

- Is the breed labour intensive?

- Find out what maintenance the breed requires etc.

- What are the breed's strengths and weaknesses?

The more knowledge you have about a particular breed, the easier it will be to plan, cost and manage a successful breeding program.

Once this initial planning is complete and all the information is on hand, the following aspects of a successful breeding program need to be considered.

c. Paddock Design:

The purpose of a well-designed paddock layout is to enable the cattle to be calmly mustered into one mob, then walk that mob from their grazing paddock to the yards to be processed. It also helps in moving cattle from one paddock to another with the minimum of stress to the animals and with safety to the cattle and stockpersons alike.

There are different designs of paddock layouts that work well, and some of the features to be considered are;

- Shade.

- Water and the distance to the water.

- Type of country.

- Quality of pasture.

- Easy access to the yards when mustering.

If a farmer is planning to build a new set of cattle yards, one good idea is to place the new yards in the middle of the grazing area or property. This would reduce the distance some of the cattle would have to travel to the yards when being mustered, compared to a set of yards that was placed at one end of the property, where parts of the grazing area would be a longer distance from the yards.

A couple of paddock layouts that do work efficiently and do not cause unnecessary stress to the cattle are laneways from the paddocks to the yards, or into smaller holding paddocks adjoining the yards. The larger grazing paddocks would then feed into this system making the cattle work flow with the minimum of stress.

Believe it or not, cattle do get to know a paddock system and do learn to work through the system with little effort or stress.

Laneways can save on labour costs, as fewer stockpersons are required to muster and move the cattle to the yards. This is very important in these times of increasing costs to cattle farmers. Laneways also give stockpersons more control over the movement of cattle.

Basic paddock laneway design.

Laneway from large grazing paddock leading into a smaller holding paddock

Spend some time on the planning of your paddock's lay out, it's very important in terms of reducing stress to the cattle when being mustered and time taken to complete the job. Once a paddock plan is built, the system will last for years so do not take short cuts.

If a property is in a high rainfall area where there is always an abundance of quick growing, high nutritional pasture available, then a system called "Cell Grazing" could be used.

Under this system, there are a number of small paddocks built around one central smaller paddock, where the water is placed. With this design, each small grazing paddock will water at the same watering point, one paddock at a time.

They will then be moved to a fresh new paddock every couple of days or whenever the requirement demands.

The idea is that the cattle are exposed to fresh pasture every couple of days without travelling a long distance from one paddock to another or to water.

Small holding paddock adjoining cattle yards.

This system can also be used to control unwanted weeds in the pasture, as they are not given the chance to grow, seed and multiply with the regular grazing. In addition, it can be argued, that a cattle farmer will become more familiar with and will get to know their cattle better, because they are with the cattle more often and observe them on more regular basis.

When designing a paddock layout and access to the yards, keep in mind the stress factor on the cattle that can be caused by the mustering.

Cattle that are less stressed;

- Have improved overall wellbeing.

- Put on condition faster and therefore are ready for market sooner.

- The cows handle well and have quieter calves as they learn from their mothers.

- Cows go back in calf earlier.

- Cattle will settle quicker when put in a new paddock and are less inclined to jump or crawl out.

- Are easier to muster, handle and manage with less of a safety risk to the animal and stockperson.

At market time the quality of the meat is more tender and saleable to consumers and therefore more profitable to the producer.

Cattle can become quiet from being WELL HANDLED, rather than from being

PAMPERED and will respond better to the control of the stockperson. This can only be made possible by the knowledge and ability of the stockpersons and the quality and design of the working infrastructure, such as paddock layouts, laneways etc. and yard design.

Whatever design is chosen will depend on the farmer's requirements and capital available. If possible, do not take short cuts with fencing, because poor fences can cause a lot of extra work when cattle keep getting out or mobs being mixed. This is extra handling and stress because the cattle are being mustered more often and extra expense is incurred in running the breeding program and property as a whole.

Poor fencing can have a negative effect on a breeding program, especially if the breeders are mated in a controlled mating program, where cows are selected to be mated to a particular bull to improve a certain breeding trait that may be a weakness in a herd.

If the different mobs cannot be kept apart with any certainty, then it is difficult to guarantee the breeding or bloodlines of the progeny from a certain bull. Therefore, the results of your long-range breeding program can be in doubt or extended. If a cattle farmer is selling stud cattle, then the pedigree of a sale animal could not be guaranteed.

Another problem that can arise due to poor quality fencing is, when maiden heifers that are being kept apart from other cattle get out or bulls get in, resulting in maiden heifers becoming pregnant too early, or to the wrong bull. Heifers becoming pregnant too early can often have stunted growth after the birth of their calf and a whole year's breeding can be wasted.

Good quality paddock fence, secure with plenty of height.

Good fences are also very important when farmers are sowing new pastures, spelling paddocks or trying to over-graze a paddock so new fresh feed will grow following the next rain. Landowners also have a legal requirement to keep their cattle off public roadways and out of their neighbour's paddocks. Only good fences will achieve this requirement.

d. Watering Points:

Watering points are not only a source of clean fresh water for cattle to drink, but can be a very useful management tool when breeding and handling cattle. Therefore, a lot of consideration is needed when planning and positioning the watering point in a paddock. Watering points can either be a running creek or river, earth dam or water trough.

Farmers who have creeks running through properties could fence along the banks then down to the water's edge at certain water holes, where the owner can then control the water access position and period of time when access is available. Those farmers who are planning their paddocks or putting in new watering points could consider the following ideas.

Because there is a wide range in property sizes, farmers will have to consider what management plan suits their purpose and requirements and how the watering point can benefit their whole cattle operations. A good starting point is to have only one watering point in each paddock. For large properties such as Cattle Stations in northern and central Australia, one watering point per grazing area could work to the owners benefit.

To achieve this, fence around each watering point so cattle can be locked off water in a particular area and forced to move and water at a different watering location when required. This can be a great help in the breeding program with the mustering procedure, as well as with pasture management.

For the purpose of this book and for the majority of the properties it will be considered that most paddocks range from one up to two thousand acres in size.

The positioning of the watering point in each paddock can either make cattle mustering a stressful, time-consuming task, or an enjoyable experience. Placing the water in the middle of the paddock can help with cattle more evenly grazing the paddock. Therefore, the pasture is not too eaten out from over grazing in certain areas, as the cattle are more spread out over the whole paddock most of the time.

It will shorten the distance cattle have to walk to water therefore less body condition is walked off the cattle during their normal daily grazing routine.

If it can be avoided, do not make cattle walk more than a couple of kilometres or miles to water, especially in the hot summer months. The idea is for cattle to graze and put on body condition and to keep it on, not put on condition and then lose it by walking long distances to water.

As young calves start to grow, they will start to walk and follow their mothers in her daily routine from the feed to water and back again.

During dry periods in large paddocks with limited watering points, problems can arise as the feed close to the watering point starts to be eaten out and the cattle, including the young calves have to walk further from the available feed to water and back again. Cattle can lose a lot of weight during this time and in severe droughts, loses can occur.

Another suggestion to think about is, what gate the cattle will be mustered out through when mustering a paddock. Once that is known then place the water a little closer to that gate.

By doing this and keeping in mind that cattle will usually work towards water when being mustered, it can make mustering easier and quicker to gather the cattle from the paddock into one mob and then move them out of the paddock.

The main reason the author has suggested that each paddock should only have one watering point, is because it could be stated that a single watering point in a paddock has more of an influence over the conception rate of a breeding program than any other factor, except the condition of the cows and the quality of the feed.

From the author's fifteen years of doing artificial breeding on beef cattle, it was observed that cycling cows tend to hang around water more when they are on heat. If there is only one watering point in a paddock then all the cycling cows will gather around that one water sometime during the day.

Earth dam to water cattle, large enough to hold water for long periods and surrounded by plenty of shade trees.

This makes the job for the bulls or the stockperson carrying out oestrus observations, much easier and a lower number of cycling cows will be missed, resulting in a higher conception rate over the herd in a shorter number of cycles and cows can then be mated over a shorter breeding season.

Less feed is then required for the breeding season if the rainfall has not been favourable.

There can be great savings on capital expenditure as fewer bulls are required, because if cycling cows are divided between two or more watering points in a paddock, then a larger percentage of bulls are required to cover the cows. This increases the cost per calf on the ground and reduces the profit margin to the farmer.

Because of the cows conceiving over a shorter period and therefore calving over a shorter period, this can help in the overall management of the breeding program, as less time is then required to observe the cows calving, which can be a time consuming job.

Calves that are born over a short calving period are more uniform in size as a mob and therefore, will sometimes sell for a premium at market.

They are uniform when kept as future breeders and because their ages are closer together, their growth rates and conformation can be more accurately assessed at selection time.

This can be a great advantage when comparing the progeny of one bull with that of another to see what bloodlines to use in future breeding programs. In addition, when selecting the better breeders on the weaning weights of their progeny, a more accurate comparison can be made between breeders and their progeny running in the same mob on the same feed and at approximately the same age.

One watering point in each paddock can save on time when the farmer is checking the cattle for ticks, lice, pinkeye, calving problems, or for any other reason. In most environments, usually about midday all cattle in a paddock are in and around the watering point, so any checking can be carried out more efficiently at that time.

Cows and calves being checked around the water at noon.

A handy hint that usually works is, for the stockperson to take a bale of hay with them when checking the cattle, one bale for 80-100 cows is enough, keeping in mind it is only to get the cattle to stay around the water while they are being checked.

The cows then also relate humans to a pleasant

experience and the calves will learn from a very early age to be calm and not stressed when being around people and vehicles.

It is a good idea to check the herd this way about once every one to two weeks, or more often during calving time, especially if first calving heifers are involved. This management practice keeps the cattle familiar with humans and keeps an eye on any problems that may arise in the herd.

There is no limit to the size of the cattle property that can benefit from a good water-management plan. On large cattle stations in central and northern Australia, paddock sizes can range anywhere from 5,000 acres up to 50,000 acres or more and as one can imagine, the mustering routine can take a lot of planning and implementing. Watering points can play a very large part in this process.

Imagine a large grazing paddock of 10,000 acres, which has four watering points in it as either, a water bore or as a large earth dam. Fence around each watering point, so in affect the fenced off watering point which has two gates, one on the opposite side to the other, ends up being a small holding paddock of only a few acres. During normal grazing routines these gates would be left open so the cattle have unlimited access to the water at all times.

In the two gateways, replace the gates with two 'TRAP GATED', which allow the cattle to pass through them from ONE DIRECTION ONLY.

The 'trap gate' is really a one-way steel gate system which allows the cattle to pass through from one direction only, then blocks the cattle from returning through the same trap gate. So that cattle can move freely to and from the watering point during normal grazing periods, there has to be two trap gates on the holding paddock fence. One on one side of the holding paddock that allows the cattle to enter the holding paddock to drink, the other one, on the other side of the holding paddock that allows the cattle to leave the holding paddock and return to their normal grazing routine.

Have the trap gates at least 100 yards or more apart so the cattle do not get confused with the "IN" and the "OUT" gates

Another great benefit of the trap gates is, that the cattle have to slow down to a steady walk to be able to walk through the gate and they can only walk through one at a time, it's a good way to steady cattle and that benefit will be noticed in other cattle handling procedures like drafting in yards etc.

This system also has other benefits to the landowner that make the idea worth serious consideration. Normally with a paddock of this size it would take a few days to muster and would need at least three or four good stockpersons to complete the job

So just imagine, the cattle are out in this large paddock and there are some cattle watering at each of the four watering points during their normal grazing period. Each of the watering points are set up with a holding paddock with two trap gates, one 'IN' gate and one 'OUT' gate. These gates can also be made to be closed from either side, so no cattle can move in or out through them.

This is how watering points can be used as a valuable management tool to aid in mustering cattle. A few days before the muster is planned to begin, drive around each of these watering points and close off the 'OUT' trap gate, but leave the 'IN' trap gate open. This way cattle can walk into the holding paddock through the

'IN' trap gate to drink, but cannot leave, so in reality the cattle end up being 'TRAPPED' in the holding paddock.

After about two days all the cattle should be trapped in the holding paddocks, so the stockpersons can take their horses or motor bikes out to the first watering point, collect the cattle that are trapped there and walk them to the next watering point and collect those cattle that are trapped and so on. Once the cattle from the last watering point are collected, walk the mob home to the cattle yards

The one watering point can supply more than one paddock as shown in the basic 'Trap Yard' design as shown on opposite page.

Basic trap yard design.

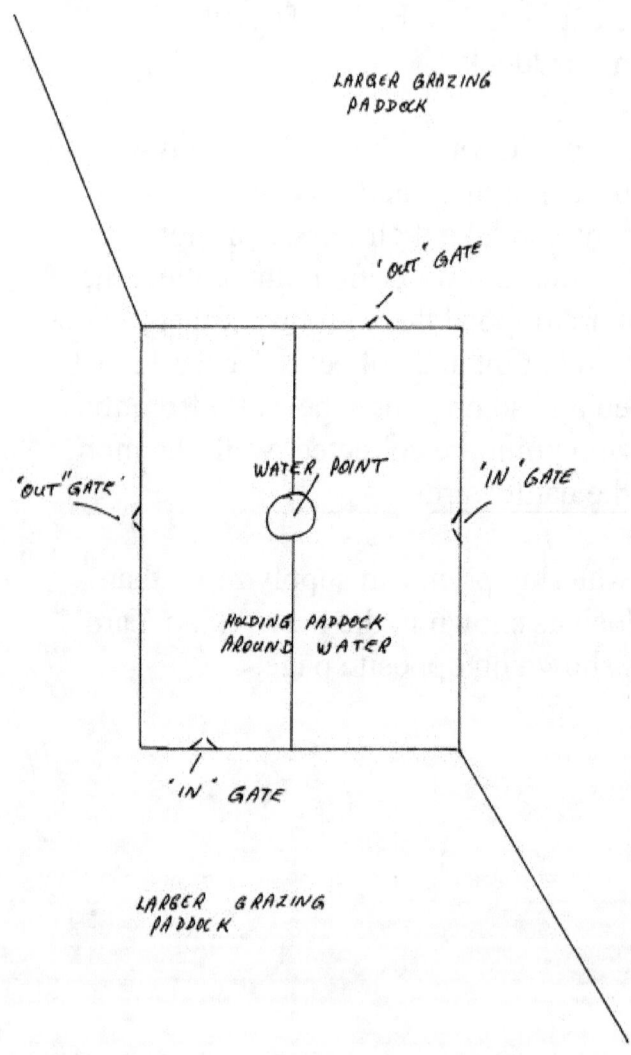

It costs money to build holding paddocks and to put up trap gates, but as one can see, it would not take long for a farmer to recover the cost of installing the trap gates from the savings on labour and time compared to a normal muster of a paddock of that size.

This system can also be used to move a mob of cattle from one paddock to another. A laneway can be built to join all the watering points into the 'one cattle movement system' ending up at the cattle yards, or into a small holding paddock adjoining the yards.

Remember, this system is mainly only used on very large cattle properties where there are large grazing paddocks, which involve large distances when mustering.

The purpose of this chapter on Watering Points is to not only consider water as a necessity to keep cattle alive, but to also point out that by using water as a management tool, a farmer's grazing system can work much more affectively, with less stress and with reduced cost outlay.

e. Pasture Requirements:

Good feed or pasture is for obvious reasons extremely important to the success or failure of a breeding program and to the growth rate and wellbeing of the cows and their progeny. To achieve a high conception rate and therefore a high calving rate from breeding, cows need to be in good body condition and on a rising plane of nutrition gaining in body weight throughout the breeding season.

This is extremely important and needs a lot of serious planning, especially for wet cows that have to produce milk for their calf at foot and to return to oestrus at the same time so they can conceive again. The plan is to have cows calving on a yearly basis.

Once calves are born, their mothers need to be able to produce enough milk to enable the calves to grow to their full genetic potential and sale target. In addition, to be well grown enough that once they are weaned, they will keep growing and gaining in body weight until they either join the breeding herd or are sold as the finished product.

Heifers that are kept for future breeders need to

be on good feed and gaining in body weight so the targeted weight at mating age can be achieved. The better the weight and overall frame size at mating, the higher the cycling rate of the heifers and therefore the higher the conception rate and calving percentage, with less chance of calving problems or stunting of the heifers' growth after they have calved.

The term 'Improved Pasture' is used a lot to describe the planting and introduction of some new species of grasses or legumes or both into a paddock, other than the natural grasses. The aim is to increase the quantity and nutritional value of that paddocks pasture. This enables the land owner to increase the stocking rate of each paddock and at the same time increase the weight gains of their cattle, which in turn increases their income.

If land owners are planning to introduce improved pasture into their paddocks they may want to consider a few suggestions first. It is not always beneficial to plant improved pasture in all types of country or in every paddock on the property in all environments.

In areas where there are usually good summer rains and dry winters, it can be found that the improved pasture will grow well in the summer months, dry off, and lose some of its nutritional value in the winter.

Problems can arise if landowners increase their stocking rate during the summer months when the feed is fresh and plentiful, then during the drier winter months cannot maintain the same quantity or quality of feed. This weather pattern will be reversed for different environments.

It has been found that in some types of country, the native grasses will hang on and maintain their nutritional value for longer than the improved pastures do during the dry periods.

By planting legumes that grow well during the winter months, like clovers and some medics, the nutritional value of the winter pasture can be increased. This is a great advantage, as the cattle will eat a combination of dry grass and fresh legume, which will help the cattle maintain or increase their body condition in the drier winter months.

Legumes are nitrogen-producing plants, which releases nitrogen-fixing bacteria into the soil from the nodules on their root systems. Besides being fresh nutritious feed for the cattle in winter, the legumes are also improving the quality of the soil as the nitrogen acts as a fertilizer, which in turn improves the quality and growth of the pasture.

Creeping Bluegrass as improved pasture.

The ideas on improved pastures can vary greatly from area to area, from one type of environment to another and between soil types.
A suggestion would be, that if land owners are planning to plant improved pasture on their property it may be an idea to contact their local

Department of Primary Industries or Department of Agriculture to get some advice on such things as:

- What improved pastures will grow best in their area in both summer and winter?

- At what time of the year does the best growth occur?

- In addition, what have the seasons been like over the past few years?

- How many good years compared to bad ones?

- How many months during the year is there good pasture growth?

- Stocking rates.

- Soil testing to find out what the soil deficiencies are

- Planting rates

- Time of the year to plant.

- Maintenance of the established pasture.

It may be a good idea to plant some paddocks down to improved pasture and keep others with native grasses, so the chance of having nutritional feed available for most of the year is increased.

The purpose of this section on pasture requirement is to impress on cattle owners, that without good quality pastures whether native or improved, it is very difficult to run a successful and profitable breeding program, in terms of growing the progeny out to their full genetic potential, which include the required mating weights or market targets.

One last point on this subject, once the pasture planning is complete and the new pasture established, it has to be maintained so that there will be good growth for years to come. To do this, always have at least one or two paddocks spelling so the pasture can grow, mature and drop seed each year or at least every couple of years. This will then thicken the pasture and will ensure future growth for years to come.

Never allow the pasture to be eaten down too

short as it will take longer to recover after rain. A good rule of thumb when planning the stocking rate of a paddock is to only graze about 50% of the bulk of the pasture each year.

Another good reason not to have pastures too short is because if the weather turns dry, the paddocks will still have a good body of dry plant roughage and in most cases, the condition and production of the cattle can still be maintained with the help of supplement feeding.

Most types of supplement feeds will only be beneficial to the cattle if there is still a good body of dry pasture available as roughage.

Native Pasture.

Also in the winter months if paddocks have a good body of ground cover in the form of dry plant roughage, the young pasture will be partly protected from the cold frosts, and in the summer months, after rain the new fresh young pasture growth will be in part protected from the hot sun.

In both cases when the pasture is made up of some fresh and some dry feed, it is a good combination of feed for cattle to gain in bodyweight. They will maintain their weight and the dry matter will stop or limit the chance of the cattle scouring from lush fresh pasture that grows after a dry period.

On the environment side of land management, erosion can be greatly reduced during heavy rains if there is plenty of ground cover from standing vegetation.

Planning paddock layouts, watering points and establishing the pasture is not the end of the property planning. Landowners have to get to know their properties as a complete working unit, how each paddock functions and what it will produce on an individual basis.

Within most properties, the soil types and ground cover can vary greatly, therefore having different levels of production at different times of the year.

Some paddocks will respond quicker than others will after rain and some will have a higher carrying capacity, others will fatten cattle quicker and the lighter country will lose some of its nutritional value in winter.

All the aspects of the property that have been discussed and the land itself are all tools to be used for cattle farmers to manage their cattle and to produce their product. For all these to work together and to function as a viable cattle-breeding property, landowners need to know their properties well. This can take two to three years of careful observation of rainfalls, weather conditions and its effect on the land, stocking rates and how quickly the land recovers after a dry period.

Record the rainfall, carrying capacity and production of each paddock, periods of the year when there is the best pasture growth with the highest level of nutritional value and how long it takes to respond after rain and the different production levels between summer and winter.

You must know this information for your property, as properties that are side by side in the same area can vary in production and viability.

Keep in mind, **'you can't manage what you don't measure'**. It is only when the landowner understand the feed, carrying capacity and weather cycles of their property, that its full potential can be utilized and the job of implementing a planned breeding program with any certainty of success, can commence.

f. Yard Design and Drafting Cattle:

A good set of cattle yards contributes greatly to the quality of the end product, which has been established, is 'Food' for human consumption. The yards should be designed in such a way as to minimise bruising and stress to the cattle when yarding up or drafting. Safety to both the cattle and operator should be of the highest priority.

The environment of the yards as a humane place of operations for the handling of cattle should be considered and steps taken to implement any infrastructure building or work that is necessary, to achieve that end.

Shade for the cattle needs to be available as either trees or fabricated structures such as roofing or shade clothe. As global warming increases, the need for animal protection from the harsh environment is just as important as that for humans.

The main receiving yard and at least one other yard need to have plenty of shade with fresh water laid on, so the cattle can cool down after being mustered before they are handled. In addition, it is a necessity to have water laid on when weaning calves, as they are in the yards for at least ten days during their education period.

All panels of the yards need to be secure and stock proof with no protruding objects, which can cause bruising and injury to the cattle. Cattle-race side panels can be covered with rubber sheeting to stop bruising and with no side vision, the cattle will then move and work more freely limiting stress. Gates need to open and close freely.

The cattle crush needs to be in good working order with all slide gates and head bail oiled and working freely, again no protruding objects. This enables cattle to be restrained securely and humanely when husbandry procedures are being conducted, with safety to the animal and operator alike. The animals should be released from the head bail as quickly as possible after the required husbandry procedure.

Square cattle yard design with high, well-built rail panels with plenty of shade from trees or man-made structures.

Design cattle yards with open clear vision for the movement of cattle from one yard to another with no harmful protruding corners or objects.

The yarding up needs to be achieved with the least amount of fuss and stress on the cattle as possible and in such a way, that cattle don't become frightened of the yards and try to avoid the yarding up in the future. This is the time when all your knowledge on cattle handling and their habits come into play and are a big part in maintaining a calm environment when working cattle in yards.

The yards should be large enough and designed in such a way as to allow the cattle to be worked freely, to minimise any stress, noise or bruising and without presenting a safety risk to the cattle or stockpersons. There needs to be enough small yards available off the drafting pen or round yard, so cattle can be drafted with ease into their selected groups, to benefit the purpose of the muster and the cattle management plan of the property. For example:

- Calves separated from their mothers.
- Age groups.
- Cattle requiring veterinary treatment.
- Sale cattle.
- Mating groups of females.
- Splitting mobs if pasture is getting short.

Round yard system with side yards for drafting cattle.

A workbench needs to be available for veterinary use and record keeping. There needs to be a roof over the crush area for shade when husbandry procedures are being carried out on individual animals, especially during artificial breeding procedures where the semen has to be protected from direct sunlight.

A safe, humane and pleasant working environment applies for the cattle and operator alike.

A waste drum should be present for hygiene purposes to dispose of any matter such as:

- Used needles
- Used drenching equipment.
- Old ear and tail tags.
- Horn tips from dehorning.
- Used cattle paint containers.
- Waste matter from desexing.
- Waste material from veterinary treatments.

Disposing of waste from around the yards can protect the cattle and stockpersons from injury and help stop the spread of any diseases that may be present.

Give the cattle a while to settle and have a drink

when first yarded before they are worked, as this will give the cows and calves a chance to mother up and relax, which makes the mob easier to handle.

Avoid having too many cattle in each yard, as smaller numbers can be worked more freely and always work the cattle in that yard from the front, the way they are required to go. This will make the cattle work in a circular movement with full vision of the stockperson at all times and if they are steadied going through the gate, will minimise bruising and stop small calves from being trampled.

Cattle crush with shade roof.

When moving cattle from one yard to another, do not just think of moving the cattle, but think 'How' to move the cattle with the least amount of effort and stress as possible.

Besides working small numbers of cattle in each yard, it is also a good idea to have a small number of stockpersons working the cattle at any one time. Depending on the size of the yards and the number of cattle being worked, one or two stockpersons are usually enough. This will enable the cattle to understand and to see where they are meant to go more easily, and will help to avoid the risk of injury to the stockperson from being knocked because of confusion by the cattle.

Cows calmly being moved from one yard to another.

Like most practical functions, it pays to be observant to see what works and what does not and to alter or change any part of the yards or the cattle working procedure that is necessary to improve the working operations in the future.

This next topic could be debated until the cows come home! However, it is doubtful if anyone can prove the principle wrong if it is given a chance to work.

If the yards are well designed and the stockpeople understand their cattle and have a good working system in place, there is no reason to use dogs or whips in the yards while working cattle for any reason.

These are both movements and sounds that will distract the cattle, they will therefore not be concentrating on the stockperson's working intentions.

Cattle are just like humans, do not frighten or harm them and they will work well and respond to your command without any stress to themselves or to you as the operator.

The safety factor when working cattle in the yards is then minimised because with dogs in the yards, cattle will often kick out as if to kick the dogs and instead, kick the stockperson working the cattle.

A piece of poly pipe can be handy for protection, or to persuade a stubborn animal to move in a different direction if need be and should be used very sparingly.

Stress and bruising to cattle through rough handling can be minimised by a well-designed set of cattle yards and the gentle approach to their work by the stockpeople, keeping in mind what the end product of their job is.

Mustering and Moving Cattle

The skill of a stockperson is not only their ability to ride a horse or a motor bike, but more importantly, it is their ability to handle and understand cattle and their habits. Although it seems a simple job to an outsider, it is a very skilful and knowledgeable job and is an extremely important part of the cattle industry.

Whether cattle are being stressed or not while being mustered, is usually the difference between a good or a bad stockperson and their attitude to the cattle and the job at hand.

The idea is to be able to understand cattle and to use this knowledge of cattle and their habits, to the advantage of the muster

Stockman mustering cows and calves in the High Country in Southern Australia during summer.

Mustering and moving cattle can be carried out on horseback, motorbikes or both and sometimes with the use of a vehicle and horse trailer to transport the stockpersons out to the back of the paddock or property. From this point, the stockpersons can then start mustering on the way back to the yards, shortening the distance they have to travel.

Only operators experienced with either horses or motorbikes should be involved in the muster for safety reasons, or at least some experienced operators should be present to make sure the muster runs smoothly. If motor vehicles and motor bikes are being used, their serviceability needs to be checked to avoid any breakdowns during the muster.

Water and lunch for the operators can be carried in the vehicle if the muster is going to take all day. Fencing equipment can be carried to repair any damaged fences or gates.

On large cattle stations, helicopters are used to locate the cattle before mustering and are often used to completely muster the paddocks and to move the cattle from the paddock to the yards or into a holding paddock adjoining the yards, where stockpersons on horseback can then take over. This way large paddocks grazing large herds of cattle can be mustered in a shorter period of time.

Two-way radios or mobile phones are sometimes used for communication between the stockpersons to check on the location of cattle and to ask for a helping hand if needed. Also by having some sort of communication between stockpersons, people can be easily located if any personal injury does occur.

The aim when mustering and moving cattle is to carry out the operation in a planned and steady manner, with the herd moving at a steady walk so as to avoid stress and also so young calves can keep up with their mothers. To achieve this, it may be necessary to stop and rest the cattle during the muster or on the way to the yards, if there is some distance to travel and the weather is hot, especially with cows and calves.

If there is a vehicle being used in the muster, it can travel along on the tail of the mob to pick up any young calves that cannot travel the distance on their own.

All personnel need to be made aware of what paddock is to be mustered, the direction of the muster, meeting locations, individual jobs and the number and type of cattle in the paddock, eg. Cows and calves, dry cows, weaners, steers or heifers etc.

By understanding cattle and their habits, stockpersons will soon observe that in most paddocks cattle will usually work to the water when they are being mustered. In addition, they get used to being moved out of the paddock through a particular gate.

Always stick to the same routine, muster a paddock the same way every time, muster to the water and move the mob out of the paddock through the same gate. Cattle will soon learn this routine and will respond to the stockperson's command and in most cases move to the water and out of the paddock with a lot less hassle and stress than would otherwise be the case.

Count the mob before leaving the paddock to make sure it was a clean muster and if not, record any deaths that may have been seen while mustering.

When mustering cows and calves it is suggested that stockpersons do not use dogs, because in most cases when the mob is being mustered and moved, the cows are usually in the lead, followed by their calves at the tail of the mob and the dogs behind them.

The dogs are normally barking or snapping at the calves and when they call out, their mothers then turn and come back to the tail of the mob. This is normal behaviour for cows that are only displaying their maternal instincts and is a trait that most cattle farmers select and breed for.

Cows turning back in the mob then cause the stockperson to push the cows and calves harder and they usually start cracking whips or yelling, which can then make the situation worse. If this happens, at the end of the day the cattle and stockpersons are worn out and the whole day's mustering has been very stressful, especially to the cattle who by now have lost some condition and are stressed.

By not being worked with dogs, the cattle will soon learn to feel at ease when being handled and mustered. They will walk out more freely and will not worry about their calves, which will just be following behind. They can then be mustered, moved to the yards and worked with the minimum of stress and in a shorter period.

Charbray cows and calves being mustered with no dogs.

In most herds, some troublesome cattle like to test the stockperson's patience by either walking off or breaking back when being mustered and driven to the yards.

To avoid this, try to read the situation and observe the mob to detect which animals are likely to be a problem. Once the problem cattle are detected, place a stockperson on the wing or in the lead or wherever else is necessary to stop the problem occurring.

With different breeds of cattle, different methods of mustering and moving the cattle need to be considered. British type breeds of cattle need to be pushed from behind, whereas Brahman type cattle like to be led by a stockperson riding in the lead, with a couple of stockpersons bring up the tail of the mob.

Hereford cows and calves being moved from paddock to yards.

It is this knowledge and understanding of one's cattle that will make a day's mustering more enjoyable to the stockpersons and less stressful to the cattle.

It is good fun to gallop across the paddock or through the scrub chasing cattle, but it is not good cattlemanship and in most cases, it can be avoided. A good cattleperson will have assessed the situation with the mob and positioned stock people around the mob in certain positions, to stop or limit the situation arising.

When the mob has arrived at the yards, let the cattle settle and rest for a while, cool down and have a drink before the drafting starts. This will let cows and calves 'mother up' and will let the cattle become aware of their surroundings.

In this book the author mentions the word '**Stress**' quite a few times and the reason being, that rough handling and a no-care attitude to the handling of cattle can, and does, cost the industry and the cattle owners a lot of money in lost production. Meat is either discarded or downgraded at slaughter. This will result in consumers turning away from eating red meat to other types of meat that they believe are more tender.

It can take cattle longer to reach their targeted market weight and can cost the cattle owner more, as additional labour is required to do the job. The bottom line to all this is, cattle owners receive a lower return for their efforts and investment.

Keeping all the topics discussed so far in mind, this book will now proceed to examine the management procedures of weaning, educating, growing, selecting, mating and calving out breeding cows, from the age of weaners.

Weaning

Using the management principle of cows calving every twelve months, weaning is the process of separating calves from their mothers and educating them for their role in life. This enables the cow time to dry off and pick up in condition before she calves again.

In most areas by now all the calves would have been marked, inoculated and ear tagged a couple of months before weaning, except on large properties where weaning and calf marking is usually done at the same time. Calf marking will be discussed later in this book.

Weaning also is the education time for young cattle, which prepares them for their future handling by stockpersons. The methods used and the quality of the weaning procedure, will govern how the weaner develops and can be handled as an animal throughout its life.

It will vary depending on the cattle, but most weanings will take about ten days to complete and is probably the most important job there is on a cattle property and should not be taken lightly or hurried.

The time to wean will depend on whether the mating season in a particular area is in spring or autumn, which is usually determined by whenever the best-feed growth is available. Most calves are weaned at about six to nine months of age.

Charbray calves ready to wean

If a cattle farmer is producing vealers and selling the calves straight off their mothers to a meat processor or through a cattle auction sale, then they will wean the calves at the older age to maximise the most growth and return. Vealers are usually sold on a monetary value per kilo or pound of body weight.

If a cattle farmer is growing the calves out to be sold at either one or two years of age, or if the heifers are to be kept as future breeders, then those calves will usually be weaned at the younger age.

Calves that are weaned younger and are grown out to be sold or mated at perhaps two years of age, are usually just as well grown and developed at that age, as those that are weaned older. The mothers of the early weaned calves will have longer to pickup in body condition before they calve again, therefore in most cases are in better shape to give birth and to produce milk to support the next calf.

Day-1

Weaning time has arrived. Muster the cows and calves into the yards and let them settle long enough for the calves to mother up and have a drink of milk. This will ensure that the calves then know where their mothers are, so if any get out for whatever reason, they will hang around the yards because that is where they know their mothers are and will not run back to the paddock.

When all the cattle are settled, draft off the calves into a yard on their own and for that day and night allow the weaner's access to water only, no hay. Calves that are to be sold as vealers can be trucked that day to the market, leaving only the calves that are to be weaned and kept, left in the yards.

For cattle farmers who breed and produce vealers only, this is the end of their job and no weaning or education of the calves in the yards is necessary. For those who breed for both the vealer market and grow out yearlings to two years old to sell on weight, or for the weaner heifers that are being kept for replacement breeders in the herd, the job of weaning is just beginning.

Now is a good time to cull some of the cows on age if required. As it is about three months since the bulls were taken out of the cows and the youngest pregnancies can be detected with a high degree of accuracy, pregnancy test all the cows.

When it is determined, which cows are empty or pregnant and their ear tag numbers recorded, make a list of the empty ones so they can be drafted off later and culled.

The aim for any breeding herd is to produce a calf every year from every cow and if a cow cannot do this, then she is not a good breeder.

Plan, select and cull by using one's head, not the heart in this breeding cow selection process.

There is no sense feeding empty cows that will not produce a return every year, a cow's role in life is to produce a calf, therefore cull empty cows so there will be more feed available for the pregnant ones. In addition, this process will increase the fertility level of the herd in the future.

A second chance may be given to young cows on their first calf if there was a dry period during the last mating. At this stage let all the cows out of the yards into a holding paddock adjoining the yards.

With the pregnant ones, now is also the time to assess the cows as breeders and their profitability as a calf producer. If the yards are equipped with a set of scales, then weigh the weaners and record their ear tag numbers and weights.

If a cattle farmer has some maiden heifers that for the first time will be going into the herd as breeders and if for example there are twenty head, then the twenty weaners with the lowest weights can be determined and drafted off.

When assessing the weaners in this way, keep in mind the age difference and make allowances if necessary.

If no scales are available, draft off the poorer quality and lighter weaners by visual observation, put them to one side and record their ear tag numbers.

Plan to purchase a set of scales for next year. The two greatest assets on cattle breeding properties and the two management tools that will provide the owner with the most information on the progress and success of their breeding program, is a set of cattle scales and the ability to pregnancy test cows.

Those twenty lighter and poorer quality weaners can then be let back into the cows to mother up, because their mothers need to be known.

Do not let all the calves out at the one time because some will mother up, have a drink and walk away and it will be very difficult to get all the tag numbers of the cows and calves. Only let two or three out at one time, mother them and record the tag numbers of the cows and calves as a pair, then let another couple out and repeat the process.

If a cattle farmer has only a few cows and the calves were ear tagged at birth or at marking time, then by now the tag numbers of the cows and their calves would have most likely been recorded, while the stockperson was carrying out one of the water and cattle checks. Cattle checks will be discussed later in this book.

Once all the cows and lighter calves are mothered and their tag numbers recorded, bring all the cows back into the yards and draft off the twenty mothered cows and calves and also the empty cows. Put the calves back with the rest of the weaners and keep the mothers of the lighter weight calves and the empty cows separate from the balance of the mob.

Those twenty cows can also be culled as reject breeders along with the empty ones, because the quality of their calves does not justify the cost and return of keeping them as breeders. The twenty replacement heifers kept from previous weanings will take their place in the herd.

The number twenty was just chosen as an example to explain the principle of selecting and culling cows by the weight and quality of their progeny at weaning time.

There is no hard and fast rule. Some cattle farmers may always cull the bottom 10% of their herd as a standard routine every year and they may have anything from 10% up to 20% of replacement heifers going into the main breeding herd.

By culling the mothers of the lighter and poorer quality weaners, this will lift the average weaning weight of the herd, which usually continues throughout the growing period of the progeny, resulting in higher body weights and a higher profit margin at time of sale.

The uniformity of the herd and progeny can also be maintained which can be a positive advantage at sale time, because buyers mostly like to see an even pen of cattle when purchasing, especially feedlot buyers.

By always keeping some replacement heifers and adding them to the main breeding herd each year, you will keep the herd young and young cows will survive the dry times more successfully.

Also new and improved genetics will be introduced into the herd more frequently, therefore lifting the breeding quality of the cows and the average weaning weight and weight gains of their progeny and any other positive breeding traits the cattle farmer has been selecting for.

Remember, although this book is mostly on the practical aspect of a breeding herd, it is also extremely important to plan, manage and implement a genetic program on improving breeding traits in a herd.

The cattle owner at this stage already knows that some of the twenty culled cows are pregnant, so either depending on their condition, feed availability or the financial position of the owner, these culled pregnant cows can be sold now or later on as a cow and calf unit when they have calved.

The purpose of culling a cow by the growth and quality of her progeny at weaning time is a good way to assess the cow as a profitable breeder. Early calving cows and therefore cows that returned to service early after the last calf will usually have the largest calves at weaning. The cows that have the smallest calves at weaning are usually the cows that were late calving or took longer to return to service last year. Either way it is desirable to cull late calving cows if a cattle farmer is trying to shorten the mating, calving periods, and the calving interval of the herd.

The value of a cow as a mother and the growth and quality of her calf is usually connected closely to the fact that the cow is a good milk producer and displays good maternal instincts. These are both traits that should and can be bred into a herd through careful visual and genetic selection.

As a rule, the quality of the calf is in most part attributed to the quality of the cow as a breeder. The quality of the weaner and how it grows and develops after weaning, is usually attributed to the animal's genes and is attributed to the sire of the calf. From this, a bull and its bloodlines can be evaluated to assess whether the owner will use that bull or bloodlines again in the future.

There are two parts to the selection process when it comes to cattle breeding, one is the cows and their genes and their quality as a breeder, the other is the bulls and their heritability factor for their positive breeding traits. Some bulls can have a positive breeding trait, but for some reason do not pass on that quality on regular basis.

At this stage a point that is worth mentioning about culling cattle, is for cattle farmers to think with their heads and not their hearts. How many times has one heard an owner say?

- "I can't sell her, her mother was a pet".

- or "I will give her another go this year".

- or "I can't sell her she's out of old Mary".

It does not matter how well a cow is bred or what her breeding history is, if she does not have a calf every year, she needs to be culled. Breeding cattle is a business and should be managed that way and all decisions should be based on a business point of view, keeping in mind the cost of production and return on capital outlaid.

Once the calves are drafted off the cows and the required culling and selection process is completed, the cows can be let out of the yards. Do not take the cows back to their paddock just yet, because next morning they will be back at the yards and all the gates and fences in between will be broken.

Just let the cows stay near the yards in a holding paddock where they will have access to feed and water for a few days, but not so close that the cows can walk right up to the yard fence and smell their calves.

The gate going back to their paddock can be left open and it will be found that after a few days the calves will stop bellowing and the cows will start to forget their calves and walk back to their own paddock without breaking any fences or gates.

Before finishing the day's work, yard up the weaners into a small secure yard where they cannot move around too much and 'rush' during the night. Besides the fact that if weaners 'rush' at night they can often break yard rails and get out, they can also injure themselves or break legs.

'Rushing' is a term used when weaners that are away from the security of their mothers, get a fright or are spooked by noises or fast movements, mainly at night and start to run blindly as one mob in any one direction. It is during this running or 'rush' that the damage to the yards or injury to the weaners occurs.

Day-2

Before letting the weaners out into a larger yard and onto water for the day, feed out hay either in hayracks or on the ground around the edge of the yard, not too much, because the weaners will still be very restless and will walk over most of the hay and some will be wasted.

Once the hay is in place let the weaners out of their night yard onto the hay and water for the day. Do not just open the gate and let all the weaners run out. The idea is to steady the weaners coming out of the night yard by walking in front of them as they walk out, talking to them at all times. This will help to steady and calm the weaners and to teach them to walk, not run and to also listen to the command and heed the control of the stockperson.

This is the beginning of the weaner's education. It will be observed by now that there is a lot of bellowing from the cows and weaners but have faith in this system as both will settle down in a few days and the weaners will start to respond to the training.

Wire-mesh hay feeding rack.

Do not do much to the weaners the second day, just let them get used to the idea of not being close to their mothers. One thing that does help is getting the weaners used to being close to humans so they will learn to trust them. Sit on the water trough and talk to the weaners, let them hear your voice and after a while it will be noted that they will start to relax and come closer to you.

When getting up to leave the yard, do so very

slowly and quietly, as on the second day the weaners will still be frightened and will be easily spooked. Each aspect of the training and education will only have a small impact on the weaners at first, but as the days go by it will be seen that the response will be more obvious and lasting on their behaviour.

At sundown yard up the weaners into the night yard again, talking to them and calming them as they are being yarded.

Day-3

Repeat the hay procedure from yesterday and let the weaners out of the night yard, walking in front of them as they go. Remember, talk to them so they will get used to the sound of the human voice.

Leave them alone for a few hours to eat the hay, as by now they are getting hungry from not drinking milk. It will take some weaners a couple of days to get use to eating hay so don't panic if some appear not to be eating much, when they do they will settle quicker and bellow less.

After lunch, the education of the weaners can

start with activities that are more direct. The idea is to gently work them through the yards from one yard to another, into the round yard or drafting pen, up the race and through the cradle, but don't catch their heads.

Every now and then, stop and block the weaners in one of the yards and hold them long enough until they stand still and settle. Keep talking to them at all times and by doing this, it is exercising your control over the weaners.

The calming sound of the stockperson's voice can be a very useful tool when handling cattle and they never seem to forget.

Keep working the weaners gently for about an hour or so then leave them alone for the rest of the day, if need be feed them more hay if they are starting to eat better. At sundown yard up the weaners into the night yard again, this in itself is good training.

On day three the weaners can be introduced onto a grain feeder if required, as this will get the weaners used to feeding from a grain feeder if they are ever placed into a feedlot.

Bulk all weather cattle Grain feeder. Day-4

The weaners by now are usually a little more settled and are getting used to being handled, are eating better and won't spook as much when the stockperson walks through the yard. There will still be a lot of bellowing from the cows and weaners, but do not take any notice of that.

On this day let the weaners out of the night yard and into the feeding yard BEFORE the hay is put out into the hay racks.

Never at any time get too casual with the way the weaners are handled or moved from one yard to another, as it won't take much to undo the effort and training of the last few days.

Once the weaners are in the feeding yard then either throw the hay over the yard fence, with the stockperson getting in the yard amongst the weaners to spread the hay around the fence, or to put it in the hayracks, or drive a vehicle into the yards and unload the hay.

The idea behind this procedure is to get the weaners used to being close to humans and vehicles while they are being fed. Weaners will then relate humans and vehicles to a pleasant experience.

When they are being checked out in the paddock in the future by the stockperson driving around the waters, they won't run away because they will have learnt that a stockperson and a vehicle is not a sound or sight to be afraid of.

Once the weaners have eaten the hay and are settled, after lunch start some more education and handling through the yards and race. Repeat the procedure from yesterday and it will be found that the weaners will respond and work more smoothly and calmly than the day before.

Please, do not try to hurry the procedure or rush away to do something else on the property. This is a very important management tool. Again, repeat this for about an hour then leave them alone for the rest of the day. At sundown, yard up the weaners into the night yard again.

Charbray weaners being educated in the yards, moving calmly from one yard to another and moved through the race and crush.

Day-5

Repeat day **4's** procedure and remember not to take any short cuts, stick to the routine, as by now the weaners will be starting to respond well to their handling. Check the cows because by now they will have started to walk out of the holding paddock, back to their original grazing paddock and should not be bellowing as much. If this is the case then move all the cows out of the holding paddock away from the yards and out of hearing distance of the weaners if possible.

If the weaners cannot hear the cows bellowing, they will settle and forget their mothers more quickly.

Observe the feeding of the weaners and if they are eating all the hay in a short period, increase the amount of hay. The more the weaners fill up on hay the more contented they will be and less likely to be still looking for their mothers.

It is not a good practice in general terms to work cattle in the yards with a horse, except on larger cattle stations where some yards are designed for cattle to be handled and drafted on horseback.

At this stage of the weaning if horses are used on the property to muster cattle, it can be helpful to use a horse in the yards to gently work the weaners. Move the weaners from one yard to another block and hold them at different times so they get used to being steadied, controlled and held in one mob by a horse.

This will get the weaners used to being handled by a horse before they are let out of the yards into an open space.

Once the mothers are away from the yards and out of hearing distance, the weaners can be left out in the water and feeding yard at night. Make sure the yards are very secure and the gates are well closed, because with any animal that is being weaned it is possible for anything to happen. Rails can be broken or gates knocked open if the weaners are still restless and the result is they will get out.

Day-6 & 7

Repeat all of day **5's** feeding and handling procedures.

Day-8

Repeat day **5** every day until the weaners are educated and settled enough to be let out into a normal grazing paddock, which will be on about day **10** or **11**.

If more stockpersons are available then on the eighth day let the weaners out into a small holding paddock where they can be 'tailed' around while feeding on grass. Do not take the letting out of the weaners into a larger paddock for the first time lightly, this procedure takes a lot of skill. (And a bit of luck).

Once the weaners are out in the open, away from the yards, which is called 'tailing', the idea is to be able to control the weaner's movements, move them as a mob and yard them up as if they were being mustered.

When letting the weaners out of the yards, have the stockpersons in the lead to steady the weaners out and to stop them from running or from breaking in all directions. At all cost try not to let the weaners run and if there is any doubt from the way they have been responding in the yards, give them another day before letting them out to be tailed.

It is at this time that all the weaning education you have been doing in the yard pays off and will make sense, of the reasons why certain management procedures are being suggested.

One idea that can help when letting the weaners out of the yards for the first time is not to feed them that day before letting them out. Instead, put some hay just outside the yard gate, then let the weaners out and hopefully they will be hungry enough to put their heads down and start eating instead of running.

As they fill up on the hay, they will start to move away from the gate slowly, grazing on the grass as they go and the stockpersons need only to steady and calm them with the sound of their voice and slow movement around the mob.

It is at this time the stockpersons need to be alert to block any weaners that try to run, remember, they have been locked up in a yard for a week. This first time out for the weaners should be in a small paddock so if some want to run they cannot go far and can be easily blocked and controlled by the stockpersons.

One thing to keep in mind, if the weaners are a bit difficult, want to run a little, are hard to control and you think you have done something wrong, you have not. Rest assured, you are not the first stockperson who has had trouble with weaners and you will not be the last. Just stay calm and remember the fundamentals, talk to the weaners a lot, do not shout, move slowly and calmly and work 'WITH' the weaners as a mob, not against them.

If all is going well tail the weaners around for an hour or so and educate them by turning the lead back into themselves now and then, keeping them in a fairly small mob and move the mob in different directions exercising control.

Let the weaners know stockmen are there and in control by talking to them at all times.

When it is thought the weaners are fairly settled and have had a good feed of grass, then proceed to yard them up by slowly moving the mob closer to the yard's receiving gate.

When the mob gets to the gate, wait until the lead weaners LOOK and SEE the gate and start walking into the yards before putting pressure on the tail of the mob to follow. Have the stockpersons keep back off the tail of the mob when yarding up, so no weaners will be forced to break back and run, but yet close enough to keep the mob moving in the one direction without turning around.

Once back in the yards, give yourself and the stockpersons a pat on the back for a job well done. As the author, it is easy to write down steps for cattle farmers to follow when weaning cattle, but the author is the first to admit that the best of plans can sometimes go wrong. Cattle farmers are dealing with animals and animals can be very unpredictable even if they are handled correctly.

If you have let the weaners out of the yards for the first time, moved them around the paddock for an hour or so and yarded them up again, without any great hassles or without losing one single weaner, then you have done well and should be very proud of yourself. It's not just a good job done today with the tailing, but for the whole of the weaning process up till now, because all the ideas and procedures talked about so far with the weaning process all lead up to this point.

If cattle farmers wean their cattle using the method and procedures as outlined in this book, they will end up with a herd of cattle that are quiet and easy to handle and muster. The cattle will do well, be more settled, will be good calf rearing mothers and the progeny will grow and develop to their full genetic potential.

If that happens, cattle farmers will have achieved their goal and the weaning process will have been a success. Once the weaners are back in the yards, secure all gates, top up the hay and leave them for the night.

Day-9

Repeat day **8** and if all is going well tail the weaners for longer.

Day-10

Repeat day **9** and if the weaners are responding well to the tailing and are settled and they know where the water is, then just ride away and leave the weaners in the holding paddock. This is now the end of the weaning for that year.

If on the tenth day it is thought the weaners need more handling before they are left out for good, then yard them up again and continue day **9** until the weaners are responding to the handling more positively.

Only when the weaners will be manageable out in paddock by themselves, let them out for good.

There is no time limit on how long a weaning should last.

Some cattle farmers may think this is a lot of work to just wean a group of calves, but over the lifetime of the animal, which in the case of a cow is somewhere between eight and twelve years, the weaning and education of the calves only takes ten days on the average. Yet will govern how that animal acts, grows, produces and is handled over the rest of its life.

If there are some older, well-behaved steers available, put them with the weaners as this will help to steady them, so when the weaners are mustered next they will be easier to handle and will learn the mustering routine from older more experienced cattle.

From now on for the purpose of this book, only the management and handling practices of the herd that include the heifer portion of the weaners will be included and discussed.

Growing and Selection Period

Before heifers can be mated, they have to reach a required frame size and body weight to cycle regularly and to be able to give birth to a calf. They need to be able to calve unassisted and produce enough milk to support the growth of the calf and yet keep growing themselves.

To achieve these requirements, careful planning needs to be given to when heifers are mated, at what age and at what weight.

First of all the heifers have to be well grown and to do this, they will need plenty of nutritional feed for the next year or so. Weaners during the next few months after weaning will tend to grow in frame rather than in body weight. This is natural and when they get to about twelve months and older they will start to put on weight and fill out in body frame.

During this growing period, a cattle breeder can weigh the heifers at intervals of two or three months to determine their weight gains and to select the heifers with the best gains to be the future breeders. When weighing, keep the weaners off feed and water for a couple of hours before each weighing. Always weigh the weaners at the same time each day as this will ensure a more accurate body weight and a more even comparison between animals.

Keep in mind, there may be more heifers than replacement heifers required to go into the main breeding herd next year, therefore the balance can be sold to other farmers or fattened and sold on weight. In addition, the weight gains of heifers from different bulls, if known, can be determined.

This can help in the selection of bulls or bloodlines for future breeding programs, with the aim to improve the growth rate of the cattle that are kept for growing and fattening purposes.

These types of cattle are usually sold based on a monetary value per kilo or pound of body weight, so the higher the body weight at sale time, the higher the monetary value per head.

It is at this time when farmers need to know which paddocks on their property are the best for growing, as discussed in the topic on Pasture

Requirements.

Once the heifers are grown to the required mating weight, the final selection should be on conformation, overall size, breed characteristics (if a cattle farmer is breeding one particular breed of cattle) and temperament. It should be kept in mind that there is a strong link between size, conformation and calving ease and a strong link between temperament and growth rates.

Conformation characteristics to look for in general are;

- A long upstanding frame, fine head features, low flank and a deep heart girth.

- Good top line with a flat tail setting or a little lowered as in the case with Brahman type cattle.

- Wide hip placement with well-developed and wide pin bones for ease of calving, witha soft loose skin and coat.

Be very critical in your judgement of the replacement heifers, there are usually more heifers available than what are needed.

Managing Natural Mating

The time to mate cows and heifers will vary between the southern and northern areas of Australia. It will depend on whether the property is in an area that has well-improved pastures and a mostly reliable rainfall, or is in the more arid areas where the native pastures are less nutritional in the winter months and the rainfall is more seasonal and is not as reliable. This reasoning will vary in different countries with different weather and seasonal patterns.

a. Mating Weight:

A rule to consider when selecting heifers to be mated is to go by the live weight of the heifers, eg. For Bos Taurus cattle (British and European Breeds), use a body weight of approx. 300 kgs. And for Bos Indicus cattle (Brahman) use a body weight of approx. 350kgs. Under normal growing conditions, when these breeds of cattle have reached those body weights they are usually mature and grown enough and are ready to be mated.

One thing that should never be overlooked is the

overall frame size of the heifers. In some areas and seasons, Bos Taurus cattle can reach those body weights at an early age, but with a smaller and less developed frame size than is required for breeding and calving out. The weight and frame needs to be considered jointly when selecting heifers to be mated.

Mature cows at mating time will, for the majority have a calf at foot, so it will have to be determined what percentage of the cows are cycling and are ready to go back in calf. Condition and body weight again are the key factors. They are an indication that the cows are cycling and are in good enough health and condition to cycle and conceive to the bull, within the required mating period.

Fat assessment of the breeding herd will allow farmers to accurately plan their management and pasture's nutritional level to ensure a high cycling rate, conception rate and calving percentage.

The following photos and body weight information has been sourced from and with the approval of, the New South Wales Department of Agriculture, NSW Australia.

The photos give a clear description of the different stages of cow's body weights in relationship to breeding requirements. From the fat score photos on pages 74 to 80, use score 2-3 as a guide to selecting breeders for mating. At these body weights, the cows should be cycling normally and should conceive early in the breeding season.

Fat scores according to the feel of an animal's fat depth.

Figure 3. Manual fat assessment sites

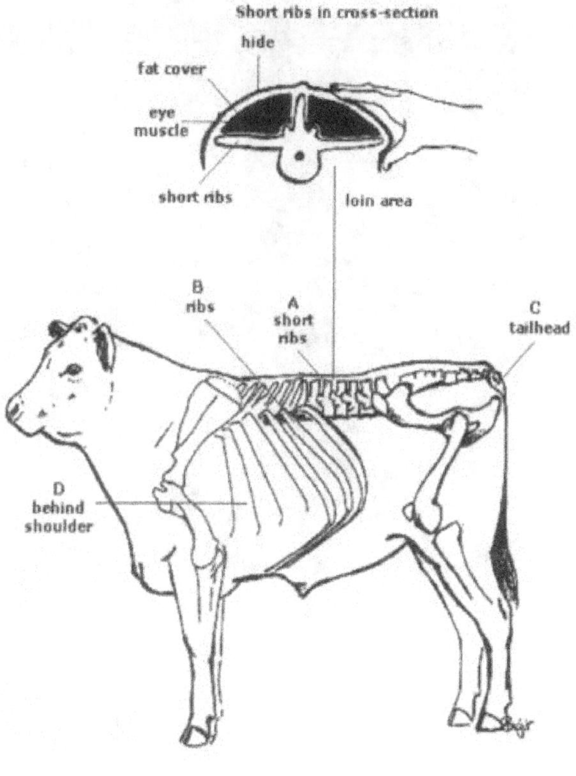

Fat score 1.

Animal is emaciated. Ribs and short ribs are sharp. There is no fat around the tailhead (C). Hip bones, tailhead and ribs are prominent.	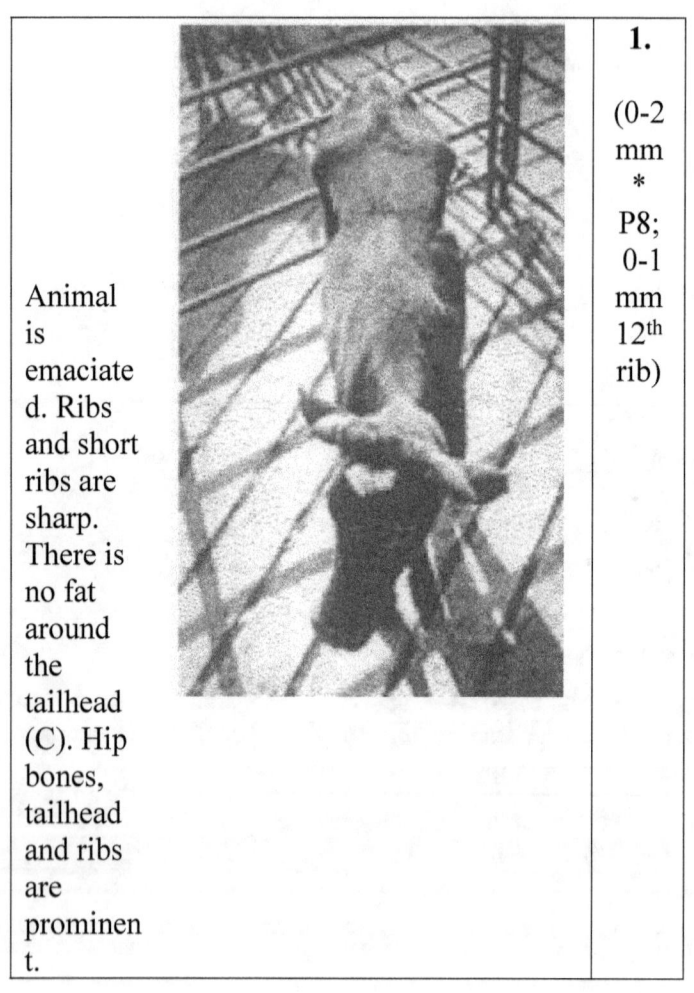	**1.** (0-2 mm * P8; 0-1 mm 12th rib)

Fat score 2.

	2. (3-6 mm P8; 2-3 mm 12th rib)
No fat beside tailhead (C). Short ribs (A) and long ribs (B) are easily distinguished. Spines feel rounded rather than sharp. Hip bone and ribs (B) are hard. Ribs are no longer visually obvious.	

Fat score 3.

3. (7-12 mm p8; 4-7 mm 12th rib)

Short ribs are prominent, rounded but still easily felt. The ribs (B) are easily felt using firm pressure to distinguish between them. Fat that is easily felt covers either side of the tailhead (C)

Fat score 4.

	**4.** (13-22 mm P8; 4-7 mm 12th rib)
Short ribs cannot be felt. There is some fat cover around the hip bone. Small mounds of fat, which are soft to touch, are present around the tailhead. Ribs are hard to feel.	

Fat score 5.

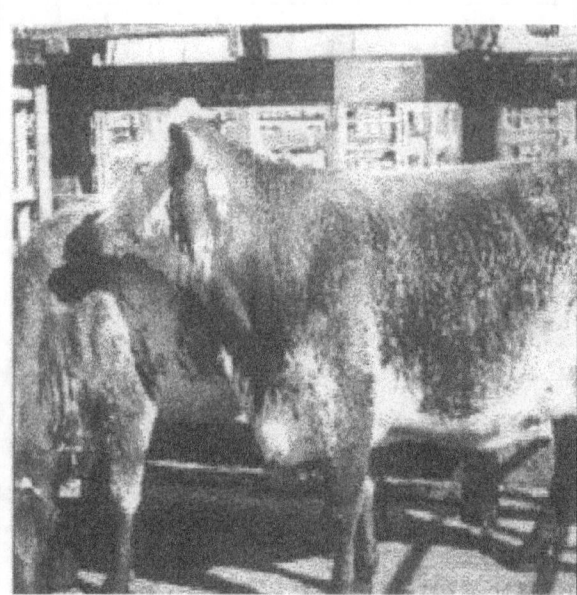

Short ribs cannot be felt. Tailhead and hip bones are almost buried in fat. Ribs (B) appear 'wavy' due to fat folds. There is fat in the brisket and udder, and squaring-off in the flank area.

5.

(23-32 mm p8; 13-18 mm 12th rib)

Fat score 6.

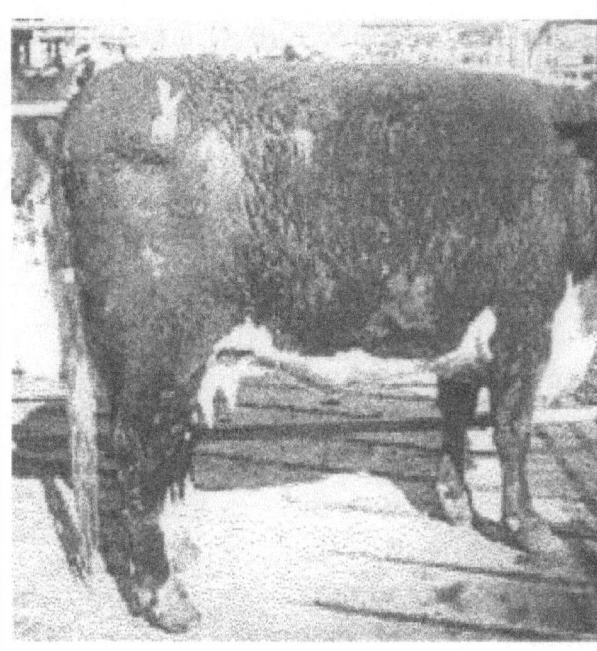

6.

(32+ mm P8; 18+ mm 12th rib)

Short ribs cannot be seen. Tailhead and hips are completely buried by large 'rounds' of fat. Ribs are 'wavy' due to fat folds. The brisket and udder are heavy. The flank is squared off and has a blocky appearance. The animal's mobility is reduced to a walk.

By manually assessing fat on your cattle, you can get a more accurate assessment, because you are actually feeling the fat deposits with the tips of your fingers.

With large herds it would take forever and a day to manually assess all the cows, but it would be possible to manually assess a small random percentage so you can get a picture in your mind of what to look for visually. That sample percentage of your herd could be assessed at different intervals prior to the mating season to determine if the cows are on the right level of nutrition and gaining in body weight prior to joining.

Prior to joining, have some paddocks spelling so there will be a good body of fresh feed available and then put the cows onto that fresh pasture at least one month before the mating season begins. This will give the cows time to get used to the pasture and to start gaining in weight and cycling ready for joining.

b. Herd Health:

Besides assessing the breeding cow's weight and condition prior to joining, look at the general health of the cows. A simple visual assessment of the cattle's health can be made by observing the following;

- Nose, mouth and tail area moist and pink with no discharge.

- Bright and alert eyes.

- Walking and grazing normally.

- Smooth and shiny coat.

- Chewing their cud while resting.

Check to see if the cows have lice, ticks, or any other external parasites and treat if necessary. A drench could be beneficial so the cows can make the most of the available pasture and use the nutrition in the pasture to increase their body weight.

If there are some calving problems in the herd or some of the cows have retained part of their after-birth after calving, contact your vet to have those cows 'flushed' before they are ready to join again. Using antibiotics in this procedure will clean the cow's uterus and reproductive organs of any uterine infections, which will ensure a quicker return to cycling and early conception when rejoined.

While the vet is on your property, ask advice on any husbandry management programs that may be implemented to improve the chances of your cows conceiving within the mating season. It may be necessary in your area to vaccinate the herd with 5in1, 7in1, or Three-Day sickness vaccine or for Botulism if your country is lacking phosphorus, but your vet will advise you of your particular needs and can take tests to verify any suspicions.

These veterinary checks should be done prior to the mating season and will cost money, but cows not going into calf will also cost money, in terms of lost income or an extended mating season with cows not calving at the right time, or at all.

c. Oestrus Observation:

Cattle farmers can make sure their cows are well fed, herd health is under control and they have a good selection program in place to select the best condition cows and heifers for mating, but it is a good idea to observe the cows cycling before joining. This will make sure the breeding program has the very best chance of being successful. Remember, every calf not born is a loss in farm income.

First of all, calculate how many cows in the herd should be cycling at the beginning of the mating season. This can be worked out by the age of the calves. During a normal season, cows will return to oestrus approximately forty-two days after calving.

If there are one hundred cows in a paddock at the beginning of the mating season (not all will have calved) and for example, going by the age of the calves, sixty cows will have calves on the ground forty-two days and older, then those sixty cows should be cycling normally if all else is right.

The average oestrus cycle of a cow is twenty-one days, so if there are a possible sixty cows that are cycling at the beginning of the season, then there are three cows cycling each day on the average. If your herd has this percentage, then that is a 100% cycling rate and the closer to this percentage your herd is cycling, the higher will be the conception at joining, the shorter will be the mating season, therefore the calving period, and the calving interval.

To check the cows cycling rate, simply observe the cows early in the morning and late in the afternoon and record the number of cows seen cycling. Repeat this procedure for about a week to get an accurate result.

The signs to observe to determine whether a cow is in oestrus or not, will include the following;

- Stands still to be mounted by another cow.
- Clear mucus discharge from the vulva.
- Swollen vulva and holding tail to one side.
- Restlessness.
- Being followed by other cows.
- Trying to mount other cows that are not on heat.
- Standing head to tail with another cow.
- In a small restless group near water.
- Bellowing calves as the cycling cow will usually hold up her milk.

- Hair rubbed over the tail head and hipbones and down the side of the flank from being mounted by other cows.

Once the health checks are completed, cows selected, paddocks with good pasture nominated, satisfied that the cows are cycling normally, then now is the time to select the maiden heifers that are to be mated for the first time.

d. Heifer's Mating Age:

Some cattle farmers will mate heifers at two years of age, others will have their heifers calving at two. The idea of calving at two years of age is that some cattle farmers believe that over the lifetime of the cow she will produce one extra calf. This is debatable because so many elements come into this system of breeding.

For example, a cattle farmer has to consider the number of heifers that may have trouble calving and sometimes with losses, when heifers are mated and calve too young. The ones that do calve all right, do they grow out as well as heifers that are mated at two years of age?

- What if the season turns dry after calving?

- Do the heifers that calve at two years really have one extra calf in their lifetime?

- At what age as cows, do those heifers stop breeding?

- Do they breed on for as long as heifers that are mated at the older age?

- Will they go back in calf easily the second time?

To mate heifers so they calve at two years, they have to be mated at fifteen months of age, which is out of timing with most main cow-mating seasons:

- When do they then fall back in line with the rest of the breeding herd?

- How long does it take?

- Are there always two mating periods?

As one can see, many questions need answers before heifers are mated. If a cattle farmer thinks through all the possibilities and the problem areas can be overcome, then mate at fifteen months.

If, in a cattle farmer's area it is possible to mate heifers at fifteen months of age, then work out a breeding selection system to select the heifers that can be mated at fifteen months. Select heifers that calve normally, return to service, conceive the second time in the required mating period and will still carry on to be good quality breeding cows in the future.

Most good breeding qualities and traits can be bred into a herd through careful selection. It is possible to breed a strain of cattle within a breed, which will produce among other qualities, heifers that will mature and can be mated at an early age, calve with minimum assistance, return to service and conceive in the required mating period and will continue to grow and develop into good quality mature cows.

But on the other hand, if a cattle farmer wants to avoid any problems that could arise with the mating of their heifers, or is unsure about how the heifers will develop prior to mating because of seasonal conditions, then stick to only one mating season a year and mate the maiden heifers at two years of age.

Bull Selection

Bull selection needs to be considered very carefully because the difference in the progeny between a good bull and a bad one can be great. The first thing to do is to evaluate the cows to determine what their faults or weaknesses are, then decide on what type of bull or bulls are needed to correct those problems. For example, sometimes it may be necessary to draft the herd up into three different groups and select three different types of bulls to correct breeding traits.

The best bull for your herd at particular time in your breeding program is not necessarily the most expensive bull on the market, or the one that may have won the grand champion prize at the local cattle show. It could just be any well-bred bull that has strong and highly heritable traits, in the areas of your herds breeding weaknesses.

Keeping in mind at all times the type of market being targeted, eg. Do you need a bull to produce vealers? Eighteen to twenty-four old fat steers or older in some cases? These are all different types of bulls.

Once a list of the herd weaknesses is known, then work out which are the most important ones, or the ones that need correcting more urgently.

If there is a long list of herd problems it is very difficult to buy a bull that is going to correct all the problems in one go, or if such a bull can be found, the improvements on each cross will be very small and it will take years to overcome any major breeding fault or weakness.

Decide on the most important weaknesses and look for a bull that has strong characteristics in one or two of those areas only. The improvement to those one or two weaknesses will be far greater and more obvious in the progeny on the first cross, than from a bull that has multiple quality characteristics.

For example;

- Use a bull this year that will improve the animal's temperament, frame and conformation.
- Then over those progeny, use a bull that will increase the milk production in the females.

- Then over those progeny use a bull that will improve weight gains and calving ease etc.

Most studs these days are in some sort of Breed Society Breed-Plan system where all the cattle registered with that Breed-Plan are evaluated and rated with a + or - breeding trait, expressed as (EBVs), Estimated Breeding Values.

Whatever breeding trait needs improving in a herd, then always select a bull that has a positive rating for that trait. When buying a bull consider the fact that a bull has approximately fifty calves a year, so a bad bull will have fifty bad calves a year, but a bad cow will have only one bad calf, therefore select bulls wisely.

These improvements can then be increased by the careful selection and culling of cows and replacements heifers as discussed in the chapters on Weaning and Growing Period.

If new bulls are to be purchased prior to joining, do so as early as possible so they will have time to adjust to their new environment and if their condition is down, it will allow time for the bulls to be fed and gain in body weight before the mating starts. Points to remember when buying bulls;

- Bulls can be checked by a vet to ensure they are healthy, possibly vaccinate for Three-Day sickness and any other disease that may be present in your area that could prevent the bull from doing his job.

- Measure the size of the bull's testes to make sure they are at least 32 cm in diameter and are normal in shape and texture.

- Check that they hang well down, not up close to the body where they can overheat and cause low fertility.

Remember, always buy cattle with a quiet and calm temperament as this trait will pass onto the progeny.

Charbray bull showing strong breed characteristics

Doing a semen test prior to joining can check semen fertility and motility.

- The number of live sperm is not as important as the number of normal and active sperm. Some bulls can have a high number of live sperm in their semen, but the motility can be poor, resulting in low fertility.

- Ask the stud owner when buying a new bull if that bull has been fertility tested, as most top studs do now as part of their sale preparations.

- Low quality semen can result in a low conception rate and low calving percentage and therefore lost income.

- Check their feet to observe any defects such as long toes or cross toes or soreness that may prevent the bull from walking correctly.

- If a bull cannot walk freely for some reason, he cannot follow and search out cycling cows.

- Drench for worms and treat for external parasites such as ticks, lice or buffalo fly.

- Trim the long hair away from the end of the bull's sheath to prevent grass seeds and dirt from collecting and causing an infection.

Brahman bull and breeding cows ready for the breeding season.

The eagerness of a bull to mate can be tested by putting the bull and a cycling cow in the yard together and observing their actions and the interest of the bull to the cow. If the bull walks away and shows no interest, then that is the way he will react when out in the paddock with cows and for obvious reasons, will not produce any or very few calves.

The cows were selected for their breeding qualities when their calves were weaned and their next calf either is on the ground or is ready to be born. The replacement heifers are grown out, the final selection made during the growing period, and their required mating weight has been reached. After careful herd assessment, bulls are selected so now is the time of year to put all that hard work and planning together and start the breeding season.

Paddock Mating System

Put the replacement heifers with the cows in the one mob or in different mobs by themselves, to suit your breeding requirements. Sometimes maiden heifers are mated separately to mature cows, with the younger bulls being joined to them, or the new bulls that are being used to correct or improve a particular breeding trait, or to avoid inbreeding. Take some time to plan the mating program, as this chapter is very important.

Depending on how many cows are in the herd, a good mating ratio is one bull to fifty cows, or three bulls to one hundred cows. If the paddocks are small eg, up to four hundred acres and there is only one watering point in the paddock, then a ratio of two bulls to one hundred cows would be enough to cover the entire herd.

As the number of cows or the number of watering points increase, so will the percentage of bulls need to increase. If the herd has different types of cows, age groups, or there are different weaknesses that need improving, or the maiden heifers are of the same bloodlines as some of the bulls, then before the bulls go out, draft the cows and heifers into the different categories and select the bulls to suit each category. Record the number in each mob and the ear tags of the cows and bulls.

Avoid if possible putting young and older bulls to work in the same paddock, as the older bulls will sometimes take over the herd and restrict the younger ones from working.

Bulls can be mated in a single sire group, or in multiple sire groups, this will depend on your breeding plan and purpose.

For example, if the mating season for the mature cows is from November 1 to January 31, a three month period, then mate the maiden heifers a month earlier on October 1 to December 31.

Next year the heifers will start and finish calving a month before the mature cows. This will give the heifers one month longer after they have calved to pick up in condition before going back into calf the second time.

The greatest stress on a breeder is having that first calf, therefore extra feed and a longer rest period is required before she cycles again and goes back in calf. By having that extra month to pickup in condition after the first calf, this can avoid some of the first calving cows from being culled as late calvers later.

For the next mating season the mature cows and the first calving cows can be mated at the same time from November 1 to January 31, and the new group of maiden heifers mated from October 1 to December 31.

Once the cows are drafted and the bulls are in their selected mobs and each mob is in its paddock and the paddock name and cattle numbers are recorded, the rest is up to nature.

Cows mated during this time will start calving from about August 5 to November 15 and the maiden heifers will start calving from July 5 to October 15. This gives the calf time to grow a little before the season breaks and there is a flush of fresh feed available from the spring rain.

Droughtmaster cows and calves during the breeding season.

In addition, the cows will not get too fat, as the pasture at the end of winter is normally eaten down and is just maintaining the condition of the cows. Most calving problems can then be avoided because the cow is not on lush feed and putting on weight during the last few months of her pregnancy, this is when the unborn calf grows the most.

The idea is to produce an average size calf at

birth and have the genetics and feed requirements for the calf to grow well after birth.

If calves are born during the lush season and their mothers are producing a lot of milk, the calf is often too small to drink all the milk and therefore, udder problems such as bottle-teats and mastitis can occur.

If in your area, the mating period is at a different time of the year, then use this system within your seasonal time frame and individual requirements.

The environments and countries can be different, but the same management procedures and principles for breeding and working cattle can apply, resulting in a smooth running breeding operation with rewarding results.

Artificial Breeding Program

If it has been decided to mate the cows using Artificial Insemination, all the cow and heifer management suggestions, especially the chapter on Oestrus Observation and all the other topics discussed in this book so far still apply, regardless if a cattle farmer is using Natural Paddock Mating or Artificial Insemination to breed their cattle.

One good thing about using artificial insemination to breed cattle is that each insemination is recorded. Records such as the cow's name or number, bull's name and date of insemination etc.

Therefore, it is known when that cow will calve so she can be observed and details such as the calving date, sex, colour, dam and sire can be recorded.

Because calves are easy to catch at the time when they are just born, some cattle farmers will catch the calves and ear tag them at this stage, so the cow and calf's identification numbers can be recorded. This information can be very useful later if cows and calves need to be mothered up for whatever reason.

With artificial breeding, it is easier to mate a certain bull to a particular cow or heifer to correct or improve genetic breeding traits, because the decision on what bull to use can be made on the spot when the assessment of a particular breeding cow or heifer can be made on an individual basis.

Artificial breeding programs are more labour intensive and time consuming than natural paddock mating, but it can be argued that with the extra time and involvement with the cattle, a cattle farmer ends up learning more about the cattle and their habits than would otherwise be the case.

This can be a good thing when young people are involved, as it gives them the chance to learn more about cattle and their habits, which in turn gives them an added advantage with more knowledge when planning future breeding programs or just mustering or working cattle in general.

The following information on setting up an Artificial Breeding Program for beef cattle is based on fifteen years of practical experience by the author, in planning and conducting beef artificial breeding programs all over Australia. Ranging from the southern areas where there are smaller properties with mostly improved pastures and a softer climatic environment, to larger properties in the north, which experience a more tropical/semi-arid environment.

Artificial breeding does work! The extent that it works on your property with your breeding cows will depend on the management of your breeding herd prior to the program commencing.

Keep in mind that if a cattle farmer uses natural paddock mating over a period of say, three months, then that is approximately four cow oestrus cycles, each cycle being approx. twenty one days on average, or four chances that the cow has to get pregnant.

With artificial breeding a program usually runs for either one or two cycles, three or six weeks, which gives the cows only two chances to get pregnant. With this in mind, it is extremely important that all the herd management aspects are considered well in advance of the program commencing, especially the chapters on oestrus observation and herd health.

Choosing a bull to use over your cows can be as easy as phoning your local Artificial Breeding Centre and asking for information on bulls of a particular breed that you wish to use. This information will usually be in the form of a photo, pedigree, progeny information such as EBV's, (Estimated Breeding Values) if the bull is part of a Breed Society Breed-Plan program, semen quality test and other information that a farmer can use in their bulls selection.

After choosing what bull to use, it is simply a matter of starting your artificial breeding program.

There are variations of systems and programs to use when conducting an artificial breeding program, but the following system will work well and will result in a high conception rate. Individual cattle farmers can modify this program to suit their own requirements.

Select the cows for the AI program and put them in a small paddock, or paddocks, close to the yards where the inseminations will take place. The idea is to have the breeders close to the yards where they can be observed for any signs of cycling every morning and late afternoon on a daily basis.

These breeding cows can be maiden heifers, mature cows or wet cows with calves at foot. A laneway joining the small paddocks to the yards will be very helpful in getting the cycling cows to the yards without any stress.

On day one of the program, start early in the morning for the first observation as cows are more active in their cycling habits early in the morning and late in the afternoons. Use either a horse or motorbike, whatever the cattle are used to, to slowly and calmly muster the cows into a mob and then let them stand and settle for a while. During this time, you can observe the cows for any signs of cycling/oestrus.

Heat detecting Charbray cows during their Oestrus Cycle.

The idea behind mustering the herd into a mob very quietly is because the signs of oestrus can be more easily detected once the feeding habit of the cows has been broken. Move slowly through the mob every now and then, take notice of anything that is different from the normal, and note the cow's identification number.

There are three different stages in a cow's

oestrus period which on average lasts thirty-six hours and a little shorter for maiden heifers.

Pre-Heat

The first stage is known as Pre-Heat, which lasts on average about eight to twelve hours. During this time, the cow will display the following signs;

- Restlessness
- Bellowing
- Cows stand head to head or head to tail
- Will attempt to mount other cows

At this stage, it is too early to inseminate the cow, as the best time to breed a cow is towards the end of the oestrus period just before ovulation, which occurs towards the end of the Post-Heat stage.

Record the cow's number but leave her out in the paddock with the rest of the cows so she can be checked again at the next observation time.

Standing-Heat

The second stage is known as Standing-Heat, which again lasts about eight to twelve hours, it's at this stage that the cow will stand still to be mounted.

During this stage, the cow will display the following signs;

- Stands still to be mounted by another cow.

- Clear mucus discharge from the vulva.

- Swollen vulva and holding tail to one side.

- Restlessness.

- Being followed by other cows.

- Trying to mount other cows that are not on heat.

- Standing head to tail with another cow.

- In a small restless group standing near water.

- Bellowing calves as the cycling cow will usually hold up her milk.

NOTE: Cows can be inseminated towards the end of this stage if necessary.

Post-Heat

The third stage is known as Post-Heat, which also lasts about eight to twelve hours and it is at the end of this stage that ovulation occurs. Signs to look for to detect any cows that may be in this last stage of their oestrus cycle are as follows;

- Hair rubbed off butt of tail or high bone.

- Dry mucus on back of legs and tail.

- Cows become settled and return to normal.

NOTE: Inseminate cows towards the end of this post heat stage for best results.

At the oestrus observation time, cows that are in standing heat in the morning can be taken back to the yards and inseminated that afternoon before the evening's observation. This timing is closer to ovulation when the egg is being released from the ovary.

If there is only one cow on heat, take a mate with her, don't leave cattle in the yards by themselves as they will get stirred up and stressed which will reduce the chances of getting the cow pregnant. Cows that are in standing heat at the evening's observation can be taken back to the yards and inseminated early next morning before the morning's observation.

In summary

Cows in standing heat in the morning--------- Inseminate that afternoon.

Cows in standing heat in the evening---------- Inseminate next morning.

For someone who is not very experienced in the heat detection of a cow in oestrus, there are different heat detection aids that can be used to help the operator detect cows in oestrus. Talk to your local Artificial Breeding Centre for advice and what type of heat detection aid will suit your requirement.

Run the AI program for about twenty-five days eg. One cycle, then again for another cycle to pick up any cow that returns to oestrus. Then wait a few days and put a bull into the herd to cover any cows that did get pregnant to the AI. From the calving dates, it can be noted what calves are the result of the artificial breeding and what calves are from the bull matings.

As indicated earlier, a one-cycle program is about three weeks and a two-cycle program is about six weeks. With hormones, it is possible to synchronise the cows that are cycling, into an eleven-day program or less, but the hormone will only work on and synchronise those cows that ARE already cycling, so the pre-checking for oestrus in your herd is extremely important.

Once the AI program is complete and the back-up bulls are in the herd, then that is the end of your artificial breeding program for that year.

Monitoring Cows from Joining to Calving

At least once a week check the breeding herd to see if any bulls have got out or have damaged themselves and are unable to service the cows. Also check to see that all bulls are working and are with the cows, not away from the mob on their own and showing no interest.

Look for problems such as external parasites that may be irritating the herd and causing them to be restless and lose weight. Sick cows that may need treatment, cows having calving problems or retaining afterbirth or any other situation like a leaking water trough or the feed getting short if there hasn't been any rain to keep the pasture growing, or any other problem that may arise that requires your attention to make the mating a success.

During the mating period keep the cows on good feed and gaining in body weight, because remember, they have to produce milk to support their calf and still have enough feed to put on weight or at least maintain their weight and cycle again. During this time there are many demands on the cows, so to give them a chance to perform, they have to be well looked after, well fed, and not mustered any more than is necessary to avoid stress.

At the end of the mating season take all the bulls out of the cows and put them back into the bull paddock until next year. Record the date when the mating season started and finished.

From this information, it can be determined when the first calf will be born, so preparations can be made and implemented in the future to cater for the calving. Preparations such as;

- Personnel to be available to assist with any calving problems that may arise.

- Have some paddocks spelling prior to calving so fresh feed will be available to put the cows and new born calves onto when calved.

- Dates can be recorded for each birth if that information is needed for a particular reason.

Once the mating season is over and all the calves from last year's mating are born, keep the herd on good feed and at least maintaining their body weight, so their calves will grow and develop right up to sale time.

Pregnant cows also need good nutritional feed to support their new unborn calf, as poor cows can sometimes abort their calves due to environmental stress, such as low quality pastures, hot weather, walking long distances to water in the summer months and external parasites.

Carrying out Calving Duties

By the time calving has arrived, the cows and heifers will most likely be in store condition or a little better under normal grazing conditions. This way most of the calving problems should be avoided, especially for the heifers.

It is known from the date when the bulls were first joined with the herd, when calving time will start and prior to that date all the equipment necessary to assist with the calving should be checked and cleaned and made workable.

 A plan needs to be in place and all stock personnel be made aware of the plan, so each stockperson knows what is expected of them and who is going to check the cows calving and what to do if problems arise.

The breeding herd should be moved closer to the yards, or a portable set of yards erected out in the paddock where the cows will be calving, so if problems do arise it will be easy to get the cow into a yard where assistance can be given as soon as possible and without causing undue stress.

Young Droughtmaster calf.

Some breeds of cattle will calve easier than others, but regardless of the breed of cattle a farmer has, it is still a good idea to check the cows and heifers calving at least once a week, or more often for some breeds, especially for first calving heifers if they are carrying too much weight, or they are young. Sometimes if the season in the previous months, leading up to calving time has been extra good and there has been lush pasture growth, calving problems could be expected as the unborn calf grows the most during the last two months of pregnancy.

While checking the cows also look for those breeders that may have calved without trouble, but will not accept their calf for some reason. Those breeders and their calves will have to be taken back to the yards in an attempt to get the cow to accept her calf and let it drink. This is rare, but it can happen.

Some cows may calve all right but in the process have been hurt internally and may not be able to walk to water. In a case like this, it may be necessary to cart the water to her, but usually the problem will right itself in a few days. If any problems arise during calving, then record the cows ear tag number and cull her at weaning time.

When observing cows calving do not be too quick to assist the cow, give her time to calve unassisted, but if after about an hour or so the calf is not born it may be necessary to assist. A mature cow that has had a few calves could be left longer before assistance was given. The idea is to work with nature, not against it.

When a calving problem is identified, try to get close enough to the cow to see whether the front or hind legs are showing. If the front legs are out, get the cow into the yards, after securing her safely in the crush, gently pull on the legs in a downward direction, and only do this when the cow herself is trying to push. Most times this is enough to help the cow calve. If not, a calf puller may be needed to get more pressure. After the calf is born let the cow out of the crush as soon as possible, so she can start licking and bonding with the calf.

If the hind legs are showing, this will most likely be a breach birth, and if no personnel on the property are experienced in assisting a cow in this condition, call your local veterinarian so the cow can be assisted as soon as possible.

Once the cow has calved just leave her and her calf in the yard with the gate left open so she can walk back to the paddock in her own time, but have access to the newborn calf until the calf is old enough to follow its mother.
Record the ear tag number of any cow that has had a calving problem or a still- born and cull

after weaning. After each assisted birth, clean all equipment and always observe hygiene protocol for the animal and personnel.

Some farmers like to ear tag the calves at birth while they are easily caught and their mother are known, then their numbers can be recorded as cow and calf, which will save the time of mothering-up later on. Birth date and sex of the calf can also be recorded at this time.

A management practice that can be beneficial during calving is, about once every three or four weeks ride through the herd, quietly cut out the cows with the oldest calves, and move them to another paddock where there is better quality feed. This then will make management a little easier as there are fewer cows still to calve in the paddock and therefore less to check on regular basis.

Also, the newborn calves of the cows that have been moved to another paddock where there is fresh feed, are given every chance to grow, develop and to reach their full genetic potential much quicker. A good start will usually result in a quicker and better quality end product. Under these conditions a more accurate assessment of the genetic value of the cows and bulls selections process, prior to the breeding season, can be validated with a higher degree of accuracy.

At the end of the calving period, record the tag numbers of the cows or heifers that have not yet calved. Even if those breeders do calve late, it could be that their gestation period is longer than normal. Therefore, their calves will usually be larger and the breeders will be prone to calving problems and should be culled. Along with any cow that was pregnant at weaning and has not yet calved, for whatever reason.

Once all the cows have calved, then move the herd into another paddock onto better quality feed, so the cows can increase their milk supply as the calves grow and develop. This quality of feed will have to be maintained throughout the calves growing stage, right up to weaning or to when the calves are sold as vealers.

As the world's seasonal conditions and weather patterns change, so will the farmer's management plans and expectations have to change, to accommodate and maintain their breeding herd to a viable level, genetically and economically.

Calf Marking

The main calf marking is usually carried out about three to six months after the first calf is born, making their age range between three and six months and one month older for the cows on their first calf. If the calves are too big at that time, then change the time to suit your requirements, or if the first calving cows are kept in a separate paddock, then it maybe more practical to have two markings.

There is no hard and fast rule on when to mark calves, work out a system that suits your management plan and requirements.

The time when the bulls are taken out of the herd can coincide with the marking of the calves, as this can save one extra muster and is sometimes used on larger cattle properties.

There are different laws and regulations in each state or country regarding the selling of cattle and how they need to be marked and identified. If cattle farmers are in doubt about the regulations that apply in their state or country, then they should contact their local Department of Agriculture before calf marking commences.

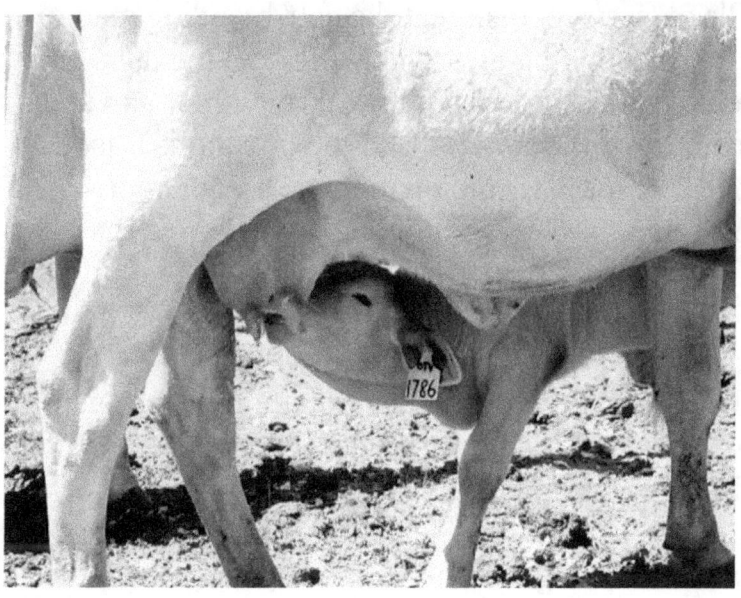

Freshly marked Charbray calf sucking 'Mum'

First job is to check all the required equipment such as, ear tag pliers, grinding stone for the desexing knife, ear marking pliers, dehorners, branding irons, and vaccination and drench guns and to make sure all are in working order and clean.

Inform the stockpersons of their duties and make sure they all know their jobs and are competent. Clean, service, disinfect the calf-branding cradle, and surrounding area, this will settle the dust. Make sure all gates are in working order and are safe to both the animal and the operators. Shade over the marking area will make the job more bearable for the operator and animal alike.

Once all the equipment is checked, muster the cows and calves into the yards and draft off all the calves. Prepare the health requirements such as antiseptic liquid or powder used when desexing the male calves and dehorning.

When marking commences, catch and secure the calf in the calf cradle in a humane and safe manner to ensure safety to both the calf and operator. Vaccinations should be administered humanely and in accordance with the manufacturer's recommendations and checking the labels for the correct dose rate.

Looking at Australia as a whole, calf marking will include some of the following procedures:

- Branding calves with a registered fire or freeze brand, also underneath the registered brand, put the year of branding, eg. For year 2002 just use the number 2. This can be useful when buying or selling cattle when the buyer wants to know the age of the cattle, which can govern the price paid.

- Ear tag the calves with their own identification number. Other information can be on the tag as well, such as the year of birth, sire number or prefix, stud information or any other information that meets the cattle farmer's requirements.

- Desex the male calves if required.

- Inoculate with 5in1 or 7in1 and any other vaccine that the area or legislation requires.

- Dehorn calves as required.

- Ear marking.

- Record the ear tag numbers and whether the calves are male or female

- Insert the NLIS (National Livestock Identification System) ear tag for lifetime animal identification.

- Record the sire of the calves or multiple sires if that is the case. This information will be vital when those heifer calves are old enough to mate and their breeding will need to be known when it comes time to select the bulls.

As soon as all the cattle procedures that relate to the calf marking time are completed, let the herd out of the yards away from the dust and dirt to protect the male calves from infection because of the desexing.

Hold the herd in a small holding paddock for a while so calves can mother up before going back to their paddock. Count the cows leaving the yards. If the herd has to travel some distance back to the paddock, move them slowly so as not to cause too much stress to the freshly marked calves, stop and rest the herd on the way if necessary.

After the completion of the marking, clean the area well and dispose of any calf residue such as skin, testes and manure off the cradle, also used drench and vaccine containers and needles. Clean and disinfect the equipment well and store until next year.

Over the next day or so, check the male calves to make sure all is well and they are grazing without any problems.

Weaning

The next management procedure is weaning time. As weaning has been discussed in detail starting on page 72, refer to that section to refresh your memory.

As it can be observed, this book has completed a full cycle on the management and handling practices of breeding cows, from the weaning of a group of heifer weaners, through their growing period and selection, their mating for the first time, calving, marking and finally when the calves from those original heifer weaners are ready to be weaned themselves.

Summary

The purpose of this book has been for the author to write down and record the knowledge and experiences gained over the years in the management and handling of beef cattle, especially breeding cows.

The aim is to pass on this knowledge to other cattle farmers, in the hope that the information might be of use to them in gaining some improvements in their current breeding programs and their cattle handling procedures in general.

The author is the first to recognize, that there are very large variations in the size of properties, in soil types, native and improved pastures, environments, weather conditions and rainfall, markets and cattle farmers requirements, beliefs and non-beliefs on how to handle cattle. However, the author believes that the basic principle of handling and breeding cattle and their daily management requirements are the same Australia wide and could be adapted successfully in other countries.

The actual breeding of the cattle is only part of the process in the line of events to produce a saleable and profitable product to consumers.

This book is not saying that cattle farmers should neglect or disregard the benefits from selecting bulls and cows with positive breeding traits and trying to improve the quality of their breeding herd by using genetics. This principle is extremely important and very necessary, but is not the only answer to improving the breeding quality of a herd.

In today's economic environment, which has touched all parts of the world, it is more important than ever to get as much gain and benefit in ones breeding program and general cattle venture as possible. It is even more important to get those gains and results with the minimum of cost and by improving or changing ones general cattle management can help greatly in this process.

For example, when selecting cattle for growth traits, it has been indicated that when measuring the difference between groups of cattle for growth, about 30% of the variation is due to genetics, and the other 70% is due to non-genetic factors.

This 70% non-genetic factor can include such things as pasture quality, breed type for a particular area and market, cattle handling, raw selection of breeding traits and general management practices.

The genetic and non-genetic percentages will vary between different breeding traits, environments and properties. The aim of this book has been to discuss some of the areas where cattle farmers can use different management and handling practices and adopt some simple selection procedures, to gain an increase in the area of the non-genetic factor.

This is often overlooked, as it is sometimes more fashionable to talk about a high priced bull or cow that a cattle farmer has just purchased, rather than look at the less expensive yet just as important, daily routine of their cattle management system which needs to work in conjunction with their herd's genetic selections.

For example, it is no good buying the best genetically bred cow if the cattle farmer takes that cow home and puts her on a barren rocky ridge and expects her to produce and give a high return.

The land itself has to be understood, how it responds in dry seasons and to what extent it recovers after rain, the type of native and improved pasture that suit the area and soil types, carrying capacity in good and bad times, whether the land will fatten or not and how each paddock performs and produces in its own right

Infrastructure needs to be in good order and functional, fences stock proof, good supply of clean fresh water in every paddock, yards well designed and built, paddock layout designed in such a way as to avoid a rodeo during mustering.

Cattle need to be well handled with a good a management system in place to breed, handle, educate, grow and select good quality cattle. A good husbandry and herd health program needs to be in place and implemented to ensure the herd is free of disease and internal and external parasites are kept under control.

Select the market that will give the best return for cattle bred and finished on your property, regardless of what the neighbour or the owner down the road produces. Properties side by side can vary in their output and production!

Keep accurate and up to date records on all aspects of the property, cattle, income returns and costs.

It is only when all these aspects of a cattle-breeding property are attended to and are functioning together, in partnership with the approach and understanding of what the end product of cattle breeding is, by the cattle farmer, that the entire cattle breeding business will operate to its full and profitable potential.